Pole-Swapping Algorithms for the Eigenvalue Problem

SIAM Spotlights is a new book series that comprises brief and enlightening books on timely topics in applied and computational mathematics and scientific computing. The books, spanning 125 pages or less, will be produced on an accelerated schedule and will be attractively priced.

Editorial Board

Peter Benner
Max Planck Institute for Dynamics of Complex Technical Systems, Magdeburg

Raymond Chan
Lingnan University, Hong Kong

Maria Emelianenko
George Mason University

Chen Greif
University of British Columbia

Jeffrey Humpherys
University of Utah School of Medicine

C. T. Kelley
North Carolina State University

Kirsten Morris
University of Waterloo

Jeffrey Sachs
Merck

Josef Málek and Zdeněk Strakoš, *Preconditioning and the Conjugate Gradient Method in the Context of Solving PDEs*

Paul G. Constantine, *Active Subspaces: Emerging Ideas for Dimension Reduction in Parameter Studies*

Dominique Orban and Mario Arioli, *Iterative Solution of Symmetric Quasi-Definite Linear Systems*

Lin Lin and Jianfeng Lu, *A Mathematical Introduction to Electronic Structure Theory*

A. Målqvist and D. Peterseim, *Numerical Homogenization by Localized Orthogonal Decomposition*

Gérard Meurant and Petr Tichý, *Error Norm Estimation in the Conjugate Gradient Algorithm*

Daan Camps, Thomas Mach, Raf Vandebril, and David S. Watkins, *Pole-Swapping Algorithms for the Eigenvalue Problem*

Pole-Swapping Algorithms for the Eigenvalue Problem

Daan Camps
Lawrence Berkeley National Laboratory, Berkeley, California

Thomas Mach
University of Potsdam, Potsdam, Germany

Raf Vandebril
KU Leuven, Leuven, Belgium

David S. Watkins
Washington State University, Pullman, Washington

Society for Industrial and Applied Mathematics
Philadelphia

Copyright © 2025 by the Society for Industrial and Applied Mathematics

10 9 8 7 6 5 4 3 2 1

All rights reserved. Printed in the United States of America. No part of this book may be reproduced, stored, or transmitted in any manner without the written permission of the publisher. For information, write to the Society for Industrial and Applied Mathematics, 3600 Market Street, 6th Floor, Philadelphia, PA 19104-2688 USA.

No warranties, express or implied, are made by the publisher, authors, and their employers that the programs contained in this volume are free of error. They should not be relied on as the sole basis to solve a problem whose incorrect solution could result in injury to person or property. If the programs are employed in such a manner, it is at the user's own risk and the publisher, authors, and their employers disclaim all liability for such misuse.

Trademarked names may be used in this book without the inclusion of a trademark symbol. These names are used in an editorial context only; no infringement of trademark is intended.

Publications Director	Kivmars H. Bowling
Executive Editor	Elizabeth Greenspan
Acquisitions Editor	Elizabeth Greenspan
Developmental Editor	Rose Kolassiba
Managing Editor	Kelly Thomas
Production Editor	Louis Primus
Copy Editor	Louis Primus
Production Manager	Rachel Ginder
Production Coordinator	Cally A. Shrader
Compositor	Cheryl Hufnagle
Graphic Designer	Doug Smock

Library of Congress Control Number: 2025008391

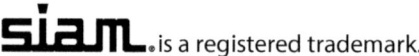 is a registered trademark.

Royalties from the sale of this book are placed in a fund to help students attend SIAM meetings and other SIAM-related activities. This fund is administered by SIAM, and qualified individuals are encouraged to write directly to SIAM for guidelines.

Contents

Preface   vii

1   Basic Facts   1
    1.1   Standard eigenvalue problem 1
    1.2   Generalized eigenvalue problem 4
    1.3   Inner products and norms 6
    1.4   Core transformations 7

2   Swapping Blocks in Triangular Matrices   13
    2.1   Single-matrix case 13
    2.2   Pencil case 21
    2.3   Backward stability in the case $m = k = 1$ 30

3   Krylov Processes   33
    3.1   Krylov subspaces 33
    3.2   The Arnoldi process 35
    3.3   Generalized Arnoldi process 37
    3.4   Rational Krylov process 38

4   Pole-Swapping Algorithms   41
    4.1   Operations on Hessenberg pairs 41
    4.2   Building an algorithm from the pieces 44
    4.3   Convergence theory 46
    4.4   Variations on the basic algorithm 51
    4.5   Aggressive early deflation 53
    4.6   Connections to earlier work 56
    4.7   Proof of Theorem 4.3.3 59

5   The Standard Eigenvalue Problem   67
    5.1   Unitary Hessenberg matrices 67
    5.2   The RQR algorithm 69

6   Krylov Processes II: Filtering   73
    6.1   Implicitly restarted Arnoldi process 73
    6.2   Filtering the rational Krylov process 76

7   Block Algorithms   79
    7.1   An algorithm that swaps blocks 79
    7.2   The Case of 2×2 Blocks 82

7.3   Bulge chasing as pole swapping 88

Bibliography   **91**

Index   **95**

Preface

This monograph is about a class of methods for solving matrix eigenvalue problems. We assume that the reader has already had some exposure to eigenvalue problems and agrees with us that they are important.

Prerequisites

The reader is expected to have some experience with matrix computations and to be familiar with the standard notation and terminology. We also assume knowledge of the basic concepts of linear algebra. The reader who mastered a big chunk of Watkins [69] or [70] will be in a good position to read this book. There are good alternatives, for example, the books by Trefethen and Bau [63], Golub and Van Loan [30], Demmel [28], and Björck [15].

This Book

The standard methods for solving the dense, nonsymmetric eigenvalue problem are descendants of Francis's implicitly shifted QR algorithm [29, 71]. If you use, for example, MATLAB or LAPACK[1] to solve dense standard or generalized eigenvalue problems, you are using a variant of Francis's algorithm. Francis and its variants are commonly known as bulge-chasing algorithms. Recently we have demonstrated that these algorithms can also be implemented by core chasing [3], which yields substantial benefits in certain cases, e.g., computation of zeros of polynomials via the companion matrix.

In this book, which is a companion to [3], not a replacement, we go in a different direction. We present a unified treatment of pole swapping, a substantial generalization of bulge chasing. The added flexibility inherent in pole swapping may allow for implementations that are substantially faster than the bulge-chasing algorithms currently in use. The jury is still out, but whether this proves true or not, the pole-swapping viewpoint is worth understanding for the substantial insights it provides into the workings of the whole class of algorithms.

As an illustration we mention that our collaborator Thijs Steel wrote the pole-swapping codes for [58]. More recently he has been engaged in improving LAPACK's QR and QZ bulge-chasing codes. QR and QZ are for the standard and generalized eigenvalue problems, respectively, and QR is expected to be twice as fast as QZ in theory. In fact the QZ code in LAPACK was, until recently, much slower than the theory predicted. Steel was able to make substantial improvements to QZ, in part incorporating insights from his experience with the pole-swapping codes. When he was done, the QZ code was faster than the theoretical prediction, relative to QR. He therefore looked into the QR code and was able to improve it as well, again using insights gained from writing the pole-swapping codes.

---

[1] MATLAB is a user-friendly proprietary package produced by the MathWorks, Inc. LAPACK [1] is an open-source project that provides high-quality software for matrix computations that is available to everybody for free. Both MATLAB and LAPACK continue to evolve. Either one can be found easily by an online search.

Some thirty years ago one of this book's authors conducted an investigation of the mechanism by which shifts are transmitted in bulge-chasing algorithms [66, 67, 68]. This resulted in concepts such as the bulge pencil, shift blurring, and bulge exchanges. All of these concepts are subsumed by the new pole-swapping theory.

Some other examples of new insights are the improved convergence theory and the discussion of the optimal packing of bulges, both in Chapter 4.

We hope you enjoy this monograph and find it to be a worthwhile read.

Chapter 1
Basic Facts

The symbols $\mathbb{R}$ and $\mathbb{C}$ denote the fields of real and complex numbers, respectively. All matrices in this book will have entries in $\mathbb{C}$. Now and then we will restrict our attention to matrices with real entries. The symbol $\mathbb{C}^{n \times m}$ denotes the set of $n \times m$ matrices with complex entries, and $\mathbb{C}^n$ denotes the set of column vectors with n complex entries (so $\mathbb{C}^n = \mathbb{C}^{n \times 1}$). We will also use the analogous symbols $\mathbb{R}^{n \times m}$ and $\mathbb{R}^n$.

1.1 ▪ Standard eigenvalue problem

Our object of study is a square matrix $A \in \mathbb{C}^{n \times n}$. The positive integer n might be large or small, but of course *large* is more challenging. A number $\lambda \in \mathbb{C}$ is called an *eigenvalue* of A if there is a nonzero vector $x \in \mathbb{C}^n$ such that
$$Ax = \lambda x.$$
The vector x is called an *eigenvector* of A *associated with* λ. Each matrix $A \in \mathbb{C}^{n \times n}$ has n eigenvalues, which are not necessarily distinct. The set of eigenvalues of A is called the *spectrum* of A and denoted $\Lambda(A)$. The basic theory of eigenvalues is covered in many books, and we will not repeat it here. We will just remind the reader about a few important facts.

Two matrices A and C are *similar* if there is a nonsingular matrix S such that
$$C = S^{-1}AS.$$
This equation is called a *similarity transformation*, and S (or S^{-1}) is called the *transforming matrix*. If A and C are similar, then they have the same eigenvalues. This suggests a methodology for computing eigenvalues: Find a similarity that transforms A to a simpler form from which the eigenvalues are evident.

A matrix $T \in \mathbb{C}^{n \times n}$ with entries t_{ij} is *upper triangular* if $t_{ij} = 0$ whenever $i > j$. Lower-triangular matrices are defined similarly. If T is upper (or lower) triangular, then its eigenvalues are exactly its main-diagonal entries $t_{11}, \ldots, t_{nn}$. The field of numerical linear algebra is biased in favor of upper-triangular (as opposed to lower-triangular) matrices. This is an historical accident, which we will, of course, perpetuate. In this book, if we speak of a *triangular matrix*, we mean upper triangular, unless otherwise stated.

If we can find a similarity transformation $T = S^{-1}AS$ for which T is triangular, this will give us the eigenvalues of A. It turns out that this is always possible in principle, and the transforming matrix can even be taken to be unitary ($S^* = S^{-1}$). This is the content of Schur's theorem [56].[2]

---
[2]Schur actually proved the theorem for lower-triangular T, the bias in favor of upper-triangular matrices having not yet been established.

Theorem 1.1.1. *(Schur) Let $A \in \mathbb{C}^{n \times n}$. Then there is a unitary $U \in \mathbb{C}^{n \times n}$ (i.e., $U^* = U^{-1}$) and a triangular T such that $T = U^{-1}AU$, or equivalently*

$$A = UTU^*. \tag{1.1.1}$$

Equation (1.1.1) is called a Schur decomposition *of A. It is far from unique; the eigenvalues of A can be made to appear in any order on the main diagonal of T.*

We omit the proof of Schur's theorem, which can be found in [70] and many other books. The one important point to be made is that the proof is not constructive: it does not yield a formula by which we can construct U and the resulting T. This is an inevitable consequence of Galois theory: There is no general formula for the eigenvalues of a matrix of dimension $n \geq 5$. Thus we cannot expect to produce the decomposition (1.1.1) all at once. Instead we must sneak up on it.

A matrix A is called *upper Hessenberg* if $a_{ij} = 0$ whenever $i > j + 1$. In the case $n = 6$, an upper Hessenberg matrix looks like this:

$$\begin{bmatrix} \times & \times & \times & \times & \times & \times \\ \times & \times & \times & \times & \times & \times \\ & \times & \times & \times & \times & \times \\ & & \times & \times & \times & \times \\ & & & \times & \times & \times \\ & & & & \times & \times \end{bmatrix}.$$

The entries marked by crosses can assume any value, while those that are left blank are zeros. Thus an upper Hessenberg matrix is *almost* upper triangular. *Lower Hessenberg* matrices are defined similarly, but we will seldom encounter them. If we refer to a *Hessenberg* matrix with no modifier, we mean *upper* Hessenberg.

Theorem 1.1.2. *Let $A \in \mathbb{C}^{n \times n}$. Then there is a unitary U and an upper Hessenberg H such that*

$$A = UHU^*.$$

This decomposition can be computed by a direct method in $O(n^3)$ arithmetic operations.

This says that every matrix is unitarily similar to an upper Hessenberg matrix, and we have a direct method (i.e., a formula) for computing U and H. The constructive proof can be found in [70] and many other places. The matrices U and H are far from unique.

Once we have a matrix in Hessenberg form, it may seem like we are almost done, but from this point on we must employ an iterative method. Each iteration is a similarity transformation that seeks to drive (some of) the subdiagonal entries $h_{j+1,j}$ toward zero. Typically none of the $h_{j+1,j}$ ever becomes exactly zero, even after many iterations, so we never get to the exact solution. In practice we never expect to get the exact solution, especially since we are working in floating-point arithmetic on a computer and making roundoff errors at every step. Throughout this book we will use the abbreviation *flop* to denote a floating-point operation. For example, the reduction to Hessenberg form requires $O(n^3)$ flops.

After a number of iterations we may make some $h_{j+1,j}$ small enough that $|h_{j+1,j}|$ is on the level of roundoff error, and we can declare it to be zero with no more harm than one single rounding operation. Once this happens, the (current) matrix has the form

$$\begin{bmatrix} H_{11} & H_{12} \\ 0 & H_{22} \end{bmatrix},$$

1.1. Standard eigenvalue problem

where H_{11} is $j \times j$. Now the eigenvalue problem splits into two smaller problems for H_{11} and H_{22}, which can be attacked separately. After repeated splittings, the matrix will be reduced to triangular form, from which we can read off the eigenvalues.

The standard methods for the iterative phase of the computation are variants of Francis's implicitly shifted QR algorithm [29]. When one speaks of the "QR algorithm," it is this procedure that is usually meant. We will call it "Francis's algorithm" to avoid confusion with the QR decomposition, which is something completely different. Detailed descriptions of Francis's algorithm can be found in [70, 71] and elsewhere. It's usually implemented as a "bulge-chasing" algorithm, but "core chasing" [3, 6] also has its advantages. Another possibility is "pole swapping," which is the subject of this book. However it is implemented, Francis's algorithm requires $O(n^2)$ flops per iteration, assuming the matrix has no special structure that can be exploited. Typically $O(n)$ iterations are needed, so the total flop count for the iterative phase is $O(n^3)$.

Francis's and other algorithms that make use of similarity transformations are useful for small- to medium-sized problems, with n not exceeding a few thousand. For larger problems, similarities are not feasible, so other methods are required. We will look at some such methods in Chapters 3 and 6. As we shall see, the large-matrix methods need the help of small-matrix methods to get the job done.

Invariant subspaces

A subspace S of $\mathbb{C}^n$ is called *invariant under A* or *A-invariant* if $AS \subseteq S$. This just means that if $x \in S$ then $Ax \in S$. The usual way to deal with a subspace is to pick a basis and work with it. Suppose S has dimension k, let $s_1, \ldots, s_k$ be a basis for S, and define an $n \times k$ matrix $S = \begin{bmatrix} s_1 & \cdots & s_k \end{bmatrix}$. We denote the *range* or *column space* of S by $\mathcal{R}(S)$. Thus $\mathcal{R}(S) = \{Sx \mid x \in \mathbb{C}^k\}$. Clearly the matrix S represents the space S in the sense that $S = \mathcal{R}(S)$. The proofs of the following simple and useful propositions are left as exercises.

Proposition 1.1.3. *Let S be a subspace of $\mathbb{C}^n$ of dimension k, and let $S_1 \in \mathbb{C}^{n \times k}$ represent S in the sense that $S = \mathcal{R}(S_1)$. Then S is A-invariant if and only if there is a matrix $C_{11} \in \mathbb{C}^{k \times k}$ such that*

$$AS_1 = S_1 C_{11}.$$

If $x \in \mathbb{C}^k$ is an eigenvector of C_{11} associated with eigenvalue λ, then $S_1 x \in \mathbb{C}^n$ is an eigenvector of A with the same eigenvalue.

Thus every eigenvalue of C_{11} is an eigenvalue of A. The eigenvalues of C_{11} are called the eigenvalues of A *associated with* the invariant subspace S.

Proposition 1.1.4. *Let S be a subspace of $\mathbb{C}^n$ of dimension k, and let $S = \begin{bmatrix} S_1 & S_2 \end{bmatrix} \in \mathbb{C}^{n \times n}$ be a nonsingular matrix whose first k columns (denoted S_1) represent S, i.e., $S = \mathcal{R}(S_1)$. Let*

$$S^{-1}AS = C = \begin{bmatrix} C_{11} & C_{12} \\ C_{21} & C_{22} \end{bmatrix},$$

where C_{11} is $k \times k$. Then S is A-invariant if and only if $C_{21} = 0$.

This proposition shows that if we have an invariant subspace in hand, we can effect a similarity transformation to block-triangular form, thereby splitting the eigenvalue problem for A into two smaller eigenvalue problems for C_{11} and C_{22}. It is fair to say that eigenvalue (and eigenvector) computation is a matter of finding invariant subspaces.

Exercises 1.1

1. Prove Proposition 1.1.3.

2. Prove Proposition 1.1.4. It is often useful, and it is certainly useful here, to express the similarity transformation in the form $AS = SC$. Then partition S and C and proceed.

3. Let A be nonsingular and let S be invariant under A. Show that S is also invariant under A^{-1}.

4. We have stated that the solution of eigenvalue problems requires iterative methods in general. This does not imply that every matrix eigenvalue problem requires iteration. For example, triangular matrices do not require iteration or any computation whatsoever. This is a trivial example. In this exercise we consider two nontrivial examples.

    (a) Show that the eigenvalues of
    $$U = \begin{bmatrix} 0 & 1 & & & \\ & 0 & 1 & & \\ & & \ddots & \ddots & \\ & & & 0 & 1 \\ 1 & & & & 0 \end{bmatrix} \in \mathbb{R}^{n \times n}$$
    are the nth roots of unity. Hint: Write out the equation $Ux = \lambda x$ in detail and find the nonzero solutions. This exercise also yields a complete set of eigenvectors.

    (b) Find the eigenvalues of
    $$T = \begin{bmatrix} 2 & -1 & & & \\ -1 & 2 & -1 & & \\ & \ddots & \ddots & \ddots & \\ & & -1 & 2 & -1 \\ & & & -1 & 2 \end{bmatrix} \in \mathbb{R}^{n \times n}.$$
    Hint: Write out the equation $Tx = \lambda x$ in detail. For each fixed λ this yields a second-order difference equation with boundary conditions $x_0 = x_{n+1} = 0$. Find those values of λ for which a nonzero solution exists. These are the eigenvalues. This exercise also yields the eigenvectors.

1.2 ▪ Generalized eigenvalue problem

Also of interest to us is the *generalized* eigenvalue problem: Given an ordered pair (A, B) of $n \times n$ matrices, a complex number λ is called an *eigenvalue* of (A, B) if there is a nonzero vector $x \in \mathbb{C}^n$ (called an *eigenvector*) such that

$$Ax = \lambda Bx.$$

If B is nonsingular, the eigenvalues of the pair (A, B) are the same as the eigenvalues of the matrices $B^{-1}A$ and AB^{-1}. We also allow infinite eigenvalues: We say that ∞ is an eigenvalue of (A, B) if 0 is an eigenvalue of the reversed pair (B, A). This happens if and only if B is singular. Clearly there is a reciprocal relationship between the eigenvalues of (A, B) and (B, A). The set of eigenvalues of (A, B) is called the *spectrum* of (A, B) and denoted $\Lambda(A, B)$.

1.2. Generalized eigenvalue problem

Instead of a pair (A, B), we may sometimes refer to a *matrix pencil* $A - \lambda B$. Whether we speak of a pair or a pencil, we are talking about the same object.

A pencil $A - \lambda B$ is called a *regular pencil* if there is at least one $\mu \in \mathbb{C}$ such that $A - \mu B$ is a nonsingular matrix. Otherwise the pencil is called *singular*. If a pencil is singular, then every λ is an eigenvalue. In this book we do not discuss singular pencils. If we use the word pencil (or pair) with no modifier, we mean *regular* pencil (or *regular* pair).

Two pencils $A - \lambda B$ and $\tilde{A} - \lambda \tilde{B}$ are called *equivalent* if there are nonsingular matrices U and V such that $\tilde{A} - \lambda \tilde{B} = U(A - \lambda B)V$, i.e., $\tilde{A} = UAV$ and $\tilde{B} = UBV$. It is easy to check that equivalent pairs have the same eigenvalues. We are mostly interested in *unitary equivalences*, i.e., equivalences for which the transforming matrices U and V are unitary.

Theorem 1.2.1. *(Generalized Schur [59]) Every pencil $A - \lambda B$ is unitarily equivalent to a pencil $S - \lambda T$, where S and T are both upper triangular.*

See, for example, [70]. One easily checks that the eigenvalues of the triangular pencil $S - \lambda T$ are s_{ii}/t_{ii}, $i = 1, \ldots, n$. Infinite eigenvalues are signaled by $t_{ii} = 0$. If, for some i, $s_{ii} = t_{ii} = 0$, the pencil is singular. Thus, if one can compute the pair (S, T) that is guaranteed by Theorem 1.2.1, one will have solved the eigenvalue problem for (A, B). However, just as for the standard eigenvalue problem, there can be no direct method to compute (S, T); iteration is required.

Just as in the single-matrix case, there is a direct method that will bring the pencil to nearly triangular form. A pencil $A - \lambda B$ is called a *Hessenberg pencil* if both A and B are upper Hessenberg. Hessenberg pencils, or Hessenberg pairs if you prefer, play a big role in this book, as we shall see, starting from Chapter 3. But for now we will focus on a special case, namely *Hessenberg-triangular form*, for which A is upper Hessenberg and B is upper triangular. Moler and Stewart [50] showed that every pencil can be reduced to Hessenberg-triangular form by a unitary equivalence.

Theorem 1.2.2. *Let $A, B \in \mathbb{C}^{n \times n}$. Then there exist unitary U and V, upper Hessenberg H, and upper-triangular T such that*

$$A - \lambda B = U(H - \lambda T)V^*.$$

This reduction can be computed by a direct method in $O(n^3)$ flops.

Moler and Stewart [50] also devised an iterative method, commonly called the *QZ algorithm*, that acts on $H - \lambda T$, reducing it to triangular form. The algorithm acts on H and T directly, but it is equivalent to Francis's algorithm applied to HT^{-1} and $T^{-1}H$ simultaneously, provided T is invertible. The original procedure chases a bulge back and forth between H and T, but it can also be implemented by core chasing [3, 4]. In Chapter 4 we will introduce a generalization that does pole swapping on Hessenberg pencils.

Deflating pairs of subspaces

Invariant subspaces play a big role in the study of the eigenvalue problem for a single matrix or operator. The analogous concept for a regular pencil $A - \lambda B$ is that of a deflating pair of subspaces. Let $\mathcal{S}$ and $\mathcal{U}$ be subspaces of $\mathbb{C}^n$ of equal dimension k. Then the pair $(\mathcal{S}, \mathcal{U})$ is called a *deflating pair* for the pencil $A - \lambda B$ if $A\mathcal{S} \subseteq \mathcal{U}$ and $B\mathcal{S} \subseteq \mathcal{U}$.

Proposition 1.2.3. *If $(\mathcal{S}, \mathcal{U})$ is a deflating pair for $A - \lambda B$ and B is nonsingular, then $\mathcal{S}$ is invariant under $B^{-1}A$, and $\mathcal{U}$ is invariant under AB^{-1}.*

Proposition 1.2.4. *Let $\mathcal{S}$ and $\mathcal{U}$ be subspaces of $\mathbb{C}^n$ of dimension k. Let $S_1, U_1 \in \mathbb{C}^{n \times k}$ be matrices such that $\mathcal{S} = \mathcal{R}(S_1)$ and $\mathcal{U} = \mathcal{R}(U_1)$. Then $(\mathcal{S}, \mathcal{U})$ is a deflating pair for the regular pencil $A - \lambda B$ if and only if there exist $C_{11}, E_{11} \in \mathbb{C}^{k \times k}$ such that*

$$AS_1 = U_1 C_{11} \quad \text{and} \quad BS_1 = U_1 E_{11}.$$

If $x \in \mathbb{C}^k$ is an eigenvector of the pair $q(C_{11}, E_{11})$ with associated eigenvalue μ, then $S_1 x \in \mathbb{C}^n$ is an eigenvector of (A, B) with the same eigenvalue.

The eigenvalues of (C_{11}, E_{11}) are called the eigenvalues of (A, B) *associated with* the pair $(\mathcal{S}, \mathcal{U})$.

Proposition 1.2.5. *Let (A, B) be a regular pair in $\mathbb{C}^{n \times n}$, let $\mathcal{S}$ and $\mathcal{U}$ be subspaces of $\mathbb{C}^n$ of dimension k, and let $S = \begin{bmatrix} S_1 & S_2 \end{bmatrix}$ and $U = \begin{bmatrix} U_1 & U_2 \end{bmatrix} \in \mathbb{C}^{n \times n}$ be nonsingular matrices such that $\mathcal{R}(S_1) = \mathcal{S}$ and $\mathcal{R}(U_1) = \mathcal{U}$. Let*

$$U^{-1}(A - \lambda B)S = C - \lambda E = \begin{bmatrix} C_{11} & C_{12} \\ C_{21} & C_{22} \end{bmatrix} - \lambda \begin{bmatrix} E_{11} & E_{12} \\ E_{21} & E_{22} \end{bmatrix},$$

where C_{11} and E_{11} are $k \times k$. Then $(\mathcal{S}, \mathcal{U})$ is a deflating pair of subspaces for (A, B) if and only if $C_{21} = E_{21} = 0$.

This proposition shows that if we can find a deflating pair of $A - \lambda B$, we can transform it to the form

$$C - \lambda E = \begin{bmatrix} C_{11} & C_{12} \\ 0 & C_{22} \end{bmatrix} - \lambda \begin{bmatrix} E_{11} & E_{12} \\ 0 & E_{22} \end{bmatrix}$$

and thereby reduce its eigenvalue problem to two smaller eigenvalue problems for the pencils $C_{11} - \lambda E_{11}$ and $C_{22} - \lambda E_{22}$.

Exercises 1.2

1. Suppose $(\mathcal{S}, \mathcal{U})$ is a deflating pair for the pencil $A - \lambda B$.

    (a) Show that if B is nonsingular, then $B\mathcal{S} = \mathcal{U}$ and $B^{-1}\mathcal{U} = \mathcal{S}$.

    (b) Prove Proposition 1.2.3.

2. Prove Proposition 1.2.4.

3. Prove Proposition 1.2.5. To this end, write the equivalence in the form $(A - \lambda B)S = U(C - \lambda E)$, i.e., $AS = UC$ and $BS = UE$, partition the matrices S, U, C, and E, and go from there.

1.3 ▪ Inner products and norms

Of all the inner products that one can define on a vector space, we will use only the standard inner product: Given $x, y \in \mathbb{C}^n$ the *inner product* of x and y is defined by

$$\langle x, y \rangle = y^* x = \sum_{i=1}^{n} x_i \overline{y}_i.$$

The vectors x and y are called *orthogonal* if $\langle x, y \rangle = 0$.

Of the many vector norms that we might consider, we will use only the Euclidean norm:
Given $x \in \mathbb{C}^n$, the *Euclidean norm* of x is defined by

$$\|x\| = \sqrt{\langle x, x \rangle} = \sqrt{\sum_{i=1}^{n} |x_i|^2}.$$

A set of vectors $v_1, v_2, \ldots, v_k$ is called *orthonormal* if the vectors are pairwise orthogonal, i.e., $\langle v_i, v_j \rangle = 0$ if $i \neq j$, and each of the vectors satisfies $\|v_i\| = 1$.

Every vector norm induces a norm on matrices. The matrix norm induced by the Euclidean vector norm is called the *spectral norm* and is defined, for any $A \in \mathbb{C}^{n \times n}$, by

$$\|A\| = \max_{x \neq 0} \frac{\|Ax\|}{\|x\|}.$$

The max is taken over all nonzero vectors in $\mathbb{C}^n$. The norms on the right-hand (defining) side of the equation are Euclidean vector norms on $\mathbb{C}^n$. The number $\|A\|$ represents the maximum magnification that a vector in $\mathbb{C}^n$ can undergo when operated on by A.

For more on inner products and norms see, for example, [70].

1.4 ▪ Core transformations

Introducing terminology that we started using a few years ago [3, 4, 5, 6, 21, 49], we define a *core transformation* (or *core* for short) to be a unitary matrix that acts only on two adjacent rows/columns, for example,

$$Q_3 = \begin{bmatrix} 1 & & & & \\ & 1 & & & \\ & & * & * & \\ & & * & * & \\ & & & & 1 \end{bmatrix},$$

where the four asterisks form a 2×2 unitary matrix. Givens rotations are examples of core transformations. Our core transformations always have subscripts that tell where the action is: Q_j acts on rows/columns j and $j+1$. The construction and use of core transformations is covered in detail in [3]. We will just present a brief introduction here.

Let us start by supposing we have an upper Hessenberg matrix A, and we wish to annihilate some or all of its subdiagonal entries. We can begin by transforming a_{21} to zero. There is a unitary matrix $\hat{Q}_1 \in \mathbb{C}^{2 \times 2}$ such that

$$\hat{Q}_1^* \begin{bmatrix} a_{11} \\ a_{21} \end{bmatrix} = \begin{bmatrix} r_{11} \\ 0 \end{bmatrix}$$

for some r_{11}. For example, we can take

$$r_{11} = \sqrt{|a_{11}|^2 + |a_{21}|^2}, \quad c_1 = a_{11}/r_{11}, \quad s_1 = a_{21}/r_{11}, \tag{1.4.1}$$

and

$$\hat{Q}_1 = \begin{bmatrix} c_1 & -\overline{s}_1 \\ s_1 & \overline{c}_1 \end{bmatrix}.$$

One easily checks that $\hat{Q}_1$ is unitary and does the job. (For a more cautious derivation of Q_1 see (2.1.16).) We also note that r_{11} is real and nonnegative. Now let $Q_1 \in \mathbb{C}^{n \times n}$ be the unitary matrix obtained by inserting $\hat{Q}_1$ into the upper left corner of an identity matrix, i.e.,

$$Q_1 = \text{diag}\{\hat{Q}_1, 1, \ldots, 1\} = \begin{bmatrix} c_1 & -\bar{s}_1 & & & \\ s_1 & \bar{c}_1 & & & \\ & & 1 & & \\ & & & \ddots & \\ & & & & 1 \end{bmatrix}.$$

If we now perform the transformation $A \to Q_1^* A$, only the first two rows are altered, and the entry a_{21} is transformed to zero.

Now that we have gotten a_{21} out of the way, we are in a position to annihilate the next subdiagonal entry, a_{32}. Let $\hat{Q}_2 \in \mathbb{C}^{2 \times 2}$ be a unitary matrix such that

$$\hat{Q}_2^* \begin{bmatrix} \tilde{a}_{22} \\ a_{32} \end{bmatrix} = \begin{bmatrix} r_{22} \\ 0 \end{bmatrix}$$

for some $r_{22} \geq 0$. Here $\tilde{a}_{22}$ denotes the $(2,2)$ entry of $Q_1^* A$, not A. $\hat{Q}_2$ can be constructed in the same way as $\hat{Q}_1$ was; we omit the details. Now we build a unitary $Q_2 \in \mathbb{C}^{n \times n}$ by inserting $\hat{Q}_2$ into an identity matrix in the right place:

$$Q_2 = \text{diag}\{1, \hat{Q}_2, 1, \ldots, 1\} = \begin{bmatrix} 1 & & & & \\ & c_2 & -\bar{s}_2 & & \\ & s_2 & \bar{c}_2 & & \\ & & & 1 & \\ & & & & \ddots \\ & & & & & 1 \end{bmatrix}.$$

Then the transformation $Q_1^* A \to Q_2^* Q_1^* A$ alters only rows two and three, creates a zero in position $(3, 2)$, and does not disturb the zero that was previously created in position $(2, 1)$. We can continue in this way if we wish, annihilating more of the subdiagonal entries.

Notation for core transformations

The matrices Q_1 and Q_2 are examples of core transformations. We have a shorthand for displaying core transformations that we have found convenient. We annihilated a_{21} above by the transformation $A \to Q_1^* A$, which transforms A by multiplying it on the left by the core transformation Q_1^*. We will depict the product $Q_1^* A$ as

$$\begin{array}{c} \\ \leftharpoondown \\ \end{array} \begin{bmatrix} \times & \times & \times & \times & \times & \times \\ \times & \times & \times & \times & \times & \times \\ & \times & \times & \times & \times & \times \\ & & \times & \times & \times & \times \\ & & & \times & \times & \times \\ & & & & \times & \times \end{bmatrix}.$$

The core Q_1^* is denoted by a small double arrow $\leftharpoondown$ that points at the first two rows of A to indicate that these are the two rows on which Q_1^* operates. The result of the operation is to transform the $(2, 1)$ entry of A to zero. Thus we have

$$\leftharpoondown \begin{bmatrix} \times & \times & \times & \times & \times & \times \\ \times & \times & \times & \times & \times & \times \\ \times & \times & \times & \times & \times & \times \\ & \times & \times & \times & \times & \times \\ & & \times & \times & \times & \times \\ & & & \times & \times & \times \\ & & & & \times & \times \end{bmatrix} = \begin{bmatrix} \times & \times & \times & \times & \times & \times \\ & \times & \times & \times & \times & \times \\ \times & \times & \times & \times & \times & \times \\ & \times & \times & \times & \times & \times \\ & & \times & \times & \times & \times \\ & & & \times & \times & \times \\ & & & & \times & \times \end{bmatrix}.$$

1.4. Core transformations

When we apply Q_2^* to Q_1^*A to annihilate a_{32}, we can depict the result as

$$\begin{bmatrix} \times\times\times\times\times\times \\ \times\times\times\times\times\times \\ \times\times\times\times\times \\ \times\times\times\times \\ \times\times\times \\ \times\times \end{bmatrix} = \begin{bmatrix} \times\times\times\times\times\times \\ \times\times\times\times\times \\ \times\times\times\times \\ \times\times\times\times \\ \times\times\times \\ \times\times \end{bmatrix}.$$

If we should continue the process by building and applying Q_3^*, we would get

$$\begin{bmatrix} \times\times\times\times\times\times \\ \times\times\times\times\times\times \\ \times\times\times\times\times \\ \times\times\times\times \\ \times\times\times \\ \times\times \end{bmatrix} = \begin{bmatrix} \times\times\times\times\times\times \\ \times\times\times\times\times \\ \times\times\times\times \\ \times\times\times \\ \times\times\times \\ \times\times \end{bmatrix}.$$

Operating with core transformations

Consider the situation we have just depicted. We want to focus on the three core transformations here, so let's just consider them separately. Writing $C_i = Q_i^*$ we have

$$C_3 C_2 C_1 = \quad .$$

It is important to note that these matrices do not commute as, in general, $C_3 C_2 \neq C_2 C_3$ and $C_2 C_1 \neq C_1 C_2$. The situation changes, however, if C_2 is left out. C_1 and C_3 do commute since the rows they act on do not overlap:

$$C_3 C_1 = C_1 C_3 = \quad = \quad = \quad .$$

In the end we have placed the symbols for C_1 and C_3 one atop the other since the order does not matter. We note that in general two core transformations C_i and C_j commute if $|i - j| \geq 2$.

Fusion

Now we consider how to deal with core transformations that act on overlapping rows and therefore interact in a nontrivial way. In the simplest case we have two adjacent core transformations acting on the same rows:

$$A_i B_i = \quad .$$

In this case we can simply multiply the two core transformations together to form a single core transformation: $A_i B_i = C_i$. Naturally enough, we call this operation *fusion*. Pictorially

$$\quad = \quad .$$

The turnover

The other major operation is the *turnover*. Suppose we have three adjacent core transformations $A_i B_{i+1} C_i$ that don't quite line up:

$$A_i B_{i+1} C_i = \text{(diagram)}.$$

If we multiply these three matrices together, the product has action only in the three rows and columns i, $i+1$, and $i+2$. It is possible to refactor this matrix in the form $\hat{A}_{i+1} \hat{B}_i \hat{C}_{i+1}$, thereby "turning over" the pattern:

$$A_i B_{i+1} C_i = \text{(diagram)} = \begin{bmatrix} \times & \times & \times \\ \times & \times & \times \\ \times & \times & \times \end{bmatrix} = \text{(diagram)} = \hat{A}_{i+1} \hat{B}_i \hat{C}_{i+1}. \quad (1.4.2)$$

This operation, which can be done in either direction, is what we call a *turnover*.

Anyone who has tried to program the turnover operation will have discovered that there are good (accurate) and bad (inaccurate) ways to do this. For a good way to do turnovers see [3], which also contains many other details. Turnover operations are crucial to our algorithm in Chapter 5.

We will briefly illustrate a typical application of the turnover. Suppose we have a product $C_1 C_2 C_3 C_4 B_2$ composed of a descending sequence $C_1 C_2 C_3 C_4$ with B_2 to the right, which can be pictured as

(diagram).

We want to somehow pass B_2 through the descending sequence. First of all, B_2 commutes with C_4, so $C_1 C_2 C_3 C_4 B_2 = C_1 C_2 C_3 B_2 C_4$. B_2 does not commute with C_3, but we can do a turnover: $C_2 C_3 B_2 = \hat{B}_3 \hat{C}_2 \hat{C}_3$, which gives $C_1 C_2 C_3 B_2 C_4 = C_1 \hat{B}_3 \hat{C}_2 \hat{C}_3 C_4$. Finally, $\hat{B}_3$ commutes with C_1, so $C_1 \hat{B}_3 \hat{C}_2 \hat{C}_3 C_4 = \hat{B}_3 C_1 \hat{C}_2 \hat{C}_3 C_4$. Putting this all together we have

$$C_1 C_2 C_3 C_4 B_2 = C_1 C_2 C_3 B_2 C_4 = C_1 \hat{B}_3 \hat{C}_2 \hat{C}_3 C_4 = \hat{B}_3 C_1 \hat{C}_2 \hat{C}_3 C_4,$$

which can be portrayed as

(diagram).

This is called a *shift-through* operation, and we will usually abbreviate it by the simple picture

(diagram).

1.4. Core transformations

The core transformation $\hat{B}_3$ that comes out on the left is, of course, different from the core B_2 that went in on the right. It's even been moved down by one position. Two of the transformations in the descending sequence have been altered as well. Since a turnover can be done in either direction, this process can be reversed, so it can be used to pass a core transformation from left to right through a descending sequence, moving the extra core transformation up by one position:

We will make use of these operations in Chapter 5.

Exercises 1.4

1. Factorize a unitary Hessenberg matrix into a sequence of core transformations. Visualize the result with the brackets with tiny arrows.

2. Let Q_i be a core transformation of the form

$$Q_i = \begin{bmatrix} c & -\bar{s} \\ s & \bar{c} \end{bmatrix}.$$

   This core transformation is a *rotation*.

   (a) Compute the determinant of Q_i.

   (b) Can all unitary matrices be factored by rotations?

   (c) How do the answers to (a) and (b) change if Q_i were a *reflection*?

3. Let $\{a, b, c\}$ be an orthogonal basis of $\mathbb{R}^3$. Show that every rotation in $\mathbb{R}^3$ can be represented as a product of a rotation around a, followed by a rotation around b, followed by a second rotation around a.

Chapter 2
Swapping Blocks in Triangular Matrices

The pole-swapping algorithms that are the subject of this book make use of processes that are exactly the same as those used to interchange two adjacent eigenvalues or blocks of eigenvalues in a triangular pencil. We will therefore study this problem in detail. We wade into the subject gradually by considering the single-matrix case first.

2.1 • Single-matrix case

Historically, the interest in swapping eigenvalues arose from the need to compute invariant subspaces. A similarity

$$A = STS^{-1}, \qquad (2.1.1)$$

with T upper triangular, reveals the eigenvalues of A on the main diagonal of T. It also yields a sequence of nested invariant subspaces. Let $\lambda_1, \ldots, \lambda_n$ denote the eigenvalues of A, listed in the order in which they appear in T, i.e., $\lambda_i = t_{ii}$, $i = 1, \ldots, n$. Assume for simplicity that the eigenvalues are distinct. Then, for each k, the first k columns of S span the invariant subspace associated with eigenvalues $\lambda_1, \ldots, \lambda_k$. This is a consequence of Proposition 1.1.3.

Now suppose we want some other invariant subspace, for example, the subspace associated with eigenvalues λ_1, λ_3, and λ_5. We cannot simply pick out the first, third, and fifth columns of S. However, if we had a method for swapping eigenvalues, e.g., a similarity transformation that swaps the positions of λ_2 and λ_3 in T, we could use that to move λ_3 into the second position. Then, by the same technique, we could swap λ_5 with λ_4 and then with λ_2 to move λ_5 into the third position. Now λ_1, λ_3, and λ_5 are in the first three positions of our updated triangular matrix T, so the space spanned by the first three columns of (the updated) S will span the corresponding invariant subspace.

If we have a method for swapping any two eigenvalues in a triangular matrix, then, starting from (2.1.1) we can produce any desired A-invariant subspace. In this chapter we will show how to do this and, more generally, how to swap two "blocks" of eigenvalues.

Swapping blocks

In the decomposition (2.1.1) the matrix T might not be triangular but only block triangular, say

$$T = \begin{bmatrix} T_{11} & \cdots & T_{1b} \\ & \ddots & \vdots \\ & & T_{bb} \end{bmatrix}, \qquad (2.1.2)$$

where the main-diagonal blocks are square and can be of varying sizes. For example, if A is real and we stay in the real number system, then it will be impossible to transform A to triangular form if any of the eigenvalues are complex. But complex eigenvalues of real matrices occur in conjugate pairs, each of which can be housed in a real 2×2 block. Thus A can be transformed to the form (2.1.2) where each block T_{ii} is either 1×1 (a real eigenvalue) or 2×2.

In general let's suppose T_{ii} has dimension k_i, $i = 1, \ldots, b$. The spectrum of A is the union of the spectra of $T_{11}, \ldots, T_{bb}$, and the decomposition $A = STS^{-1}$ also yields some invariant subspaces. (Assume that the submatrices have pairwise disjoint spectra.) The first k_1 columns of S span the A-invariant subspace associated with $\Lambda(T_{11})$, the first $k_1 + k_2$ columns span the invariant subspace associated with $\Lambda(T_{11}) \cup \Lambda(T_{22})$, and so on. But suppose we want the invariant subspace associated with just $\Lambda(T_{22})$. If we can swap the blocks T_{11} and T_{22} by a similarity transformation, then the first k_2 columns of (the transformed) S will span the desired A-invariant subspace.

Let us consider, therefore, the following problem. Given a block-triangular matrix

$$T = \begin{bmatrix} T_{11} & T_{12} \\ & T_{22} \end{bmatrix},$$

where T_{11} and T_{22} are $k \times k$ and $m \times m$, respectively, perform a unitary similarity transformation

$$\widehat{T} = Q^{-1}TQ$$

such that

$$\widehat{T} = \begin{bmatrix} \widehat{T}_{11} & \widehat{T}_{12} \\ & \widehat{T}_{22} \end{bmatrix},$$

$\widehat{T}_{11}$ is $m \times m$, $\Lambda(\widehat{T}_{11}) = \Lambda(T_{22})$, and $\Lambda(\widehat{T}_{22}) = \Lambda(T_{11})$. Typically k and m will be fairly small, and T will be a submatrix of a much larger matrix in which we are swapping two of the many blocks. We will assume that $\Lambda(T_{11})$ and $\Lambda(T_{22})$ are disjoint.

For starters a similarity transformation by

$$\begin{bmatrix} 0 & I_k \\ I_m & 0 \end{bmatrix}$$

transforms T to

$$\begin{bmatrix} T_{22} & \\ T_{12} & T_{11} \end{bmatrix}, \tag{2.1.3}$$

which accomplishes the desired swap, except that the block T_{12} is in the wrong place. We need to get rid of it. To this end we seek $X \in \mathbb{C}^{k \times m}$ such that

$$\begin{bmatrix} I_m & \\ X & I_k \end{bmatrix}^{-1} \begin{bmatrix} T_{22} & \\ T_{12} & T_{11} \end{bmatrix} \begin{bmatrix} I_m & \\ X & I_k \end{bmatrix} = \begin{bmatrix} T_{22} & \\ & T_{11} \end{bmatrix}. \tag{2.1.4}$$

This equation is equivalent to

$$\begin{bmatrix} T_{22} & \\ T_{12} & T_{11} \end{bmatrix} \begin{bmatrix} I_m \\ X \end{bmatrix} = \begin{bmatrix} I_m \\ X \end{bmatrix} T_{22},$$

which can also be written in the original (unflipped) coordinate system as

$$\begin{bmatrix} T_{11} & T_{12} \\ & T_{22} \end{bmatrix} \begin{bmatrix} X \\ I_m \end{bmatrix} = \begin{bmatrix} X \\ I_m \end{bmatrix} T_{22}. \tag{2.1.5}$$

2.1. Single-matrix case

This shows that we are looking for an X such that the columns of $\begin{bmatrix} X \\ I_m \end{bmatrix}$ span the T-invariant subspace associated with $\Lambda(T_{22})$. Equation (2.1.5) is satisfied if and only if the *Sylvester equation*

$$XT_{22} - T_{11}X = T_{12} \tag{2.1.6}$$

holds. This is a system of mk linear equations in mk unknowns that has a unique solution if $\Lambda(T_{11})$ and $\Lambda(T_{22})$ are disjoint (Exercise 2). We can solve the system stably by some variant of Gaussian elimination; the cost is negligible if m and k are small.[3]

Once we have solved for X, we can let

$$S = \begin{bmatrix} & I_k \\ I_m & \end{bmatrix} \begin{bmatrix} I_m & \\ X & I_k \end{bmatrix} = \begin{bmatrix} X & I_k \\ I_m & \end{bmatrix},$$

and note that

$$S^{-1} \begin{bmatrix} T_{11} & T_{12} \\ & T_{22} \end{bmatrix} S = \begin{bmatrix} T_{22} & \\ & T_{11} \end{bmatrix}.$$

Thus S accomplishes the desired swap, but S is not unitary. Since a unitary transformation is wanted for numerical stability, an additional step is needed, namely a QR decomposition. Let

$$\begin{bmatrix} X \\ I_m \end{bmatrix} = QR = \begin{bmatrix} Q_1 & Q_2 \end{bmatrix} \begin{bmatrix} R_{11} \\ 0 \end{bmatrix} = Q_1 R_{11}, \tag{2.1.7}$$

where Q is unitary. Since the columns of Q_1 span the same space as $\begin{bmatrix} X \\ I_m \end{bmatrix}$, namely the T-invariant subspace associated with $\Lambda(T_{22})$,

$$Q^*TQ = \begin{bmatrix} \widehat{T}_{11} & \widehat{T}_{12} \\ & \widehat{T}_{22} \end{bmatrix} \tag{2.1.8}$$

with $\Lambda(\widehat{T}_{11}) = \Lambda(T_{22})$. This is the desired transformation.

To summarize, the block swapping procedure consists of three steps:

1. Solve the Sylvester equation (2.1.6) for X.
2. Perform the QR decomposition (2.1.7) to obtain unitary Q.
3. Execute the similarity transformation (2.1.8).

Refinement

Since we are computing in floating-point arithmetic and making roundoff errors on every step, the transformation (2.1.8) will produce a matrix that is not quite block triangular. Instead it will have the form

$$\begin{bmatrix} \widehat{T}_{11} & \widehat{T}_{12} \\ \widehat{T}_{21} & \widehat{T}_{22} \end{bmatrix}, \tag{2.1.9}$$

where $\widehat{T}_{21}$ is nonzero due to roundoff errors. Normally $\|\widehat{T}_{21}\|$ will be tiny. If it is small enough, say, on the level of roundoff error relative to $\|T\|$, we can simply set $\widehat{T}_{21}$ to zero and move on.

In this section we describe a refinement procedure that can be used to drive $\widehat{T}_{21}$ to zero in cases when it is not sufficiently small. For simplicity we will drop the hats from (2.1.9) and consider

$$T = \begin{bmatrix} T_{11} & T_{12} \\ T_{21} & T_{22} \end{bmatrix}$$

---

[3]For larger k and m there are more efficient methods [11], [69, §4.8].

with $\|T_{21}\|$ small but not small enough, and look for a similarity transform that perturbs T slightly, setting T_{21} to zero. We consider first a transformation of the same type as in (2.1.4); we seek X with small norm such that

$$\begin{bmatrix} I_m & \\ X & I_k \end{bmatrix}^{-1} \begin{bmatrix} T_{11} & T_{12} \\ T_{21} & T_{22} \end{bmatrix} \begin{bmatrix} I_m & \\ X & I_k \end{bmatrix} = \begin{bmatrix} \widetilde{T}_{11} & \widetilde{T}_{12} \\ 0 & \widetilde{T}_{22} \end{bmatrix}.$$

An equivalent equation is

$$\begin{bmatrix} T_{11} & T_{12} \\ T_{21} & T_{22} \end{bmatrix} \begin{bmatrix} I_m \\ X \end{bmatrix} = \begin{bmatrix} I_m \\ X \end{bmatrix} \widetilde{T}_{11}, \qquad (2.1.10)$$

which shows that we are looking for an X with small norm such that the columns of $\begin{bmatrix} I_m \\ X \end{bmatrix}$ span a T-invariant subspace. If this equation holds for some small X, then necessarily $\widetilde{T}_{11} = T_{11} + T_{12}X$, so $\widetilde{T}_{11}$ is a slight perturbation of T_{11}, and we have computed the invariant subspace associated with $\Lambda(\widetilde{T}_{11})$. Combining the two matrix equations implied by (2.1.10), we find that (2.1.10) holds if and only if X satisfies the *algebraic Riccati equation*

$$XT_{11} - T_{22}X = T_{21} - XT_{12}X. \qquad (2.1.11)$$

This is a nonlinear matrix equation, in contrast to the Sylvester equation (2.1.6), which is linear. One can prove by standard arguments using the contraction mapping principle [59, 60], [69, §2.7] that (2.1.11) has a unique small-norm solution if $\|T_{21}\|$ is sufficiently small. By this we mean that, although (2.1.11) typically has many solutions, there is exactly one for which $\|X\|$ is small, and this is the solution that is of interest to us.

For small enough X, the quadratic term $XT_{12}X$ in (2.1.11) will be negligible in size. Therefore we can approximate X well by dropping the quadratic term and solving the resulting Sylvester equation

$$XT_{11} - T_{22}X = T_{21}, \qquad (2.1.12)$$

which is of exactly the same form (after adjustment of subscripts) as (2.1.6). Once we have solved this equation, the resulting X is not exactly the one we want, but it will be very close. Notice that this linear solve amounts to one step of Newton's method applied to the quadratic equation (2.1.11) with initial guess $X^{(0)} = 0$. Applying Newton's method to solving Riccati equations is a popular technique often referred to as Newton–Kleinman iteration [40, 14].

Once we have X, we proceed as in the original swapping step and compute a QR decomposition

$$\begin{bmatrix} I_m \\ X \end{bmatrix} = QR = \begin{bmatrix} Q_1 & Q_2 \end{bmatrix} \begin{bmatrix} R_{11} \\ 0 \end{bmatrix} = Q_1 R_{11}, \qquad (2.1.13)$$

then make the similarity transformation

$$Q^*TQ = \begin{bmatrix} \widehat{T}_{11} & \widehat{T}_{12} \\ \widehat{T}_{21} & \widehat{T}_{22} \end{bmatrix}. \qquad (2.1.14)$$

If we had obtained X by solving (2.1.11) exactly, and if we had made no roundoff errors anywhere in the computation, the columns of $\begin{bmatrix} I_m \\ X \end{bmatrix}$ would span exactly a T-invariant subspace (2.1.10), and so would the columns of Q_1 in (2.1.13). Therefore the similarity (2.1.14) would yield $\widehat{T}_{21} = 0$.

Since we do not solve (2.1.11) exactly, and we do make further roundoff errors, $\widehat{T}_{21}$ will not be exactly zero. If $\|\widehat{T}_{21}\|$ is small enough, we can set it to zero and move on. If not, we can repeat the correction step. Since Newton's method converges quadratically, very few corrections will be necessary. In fact, one correction will be enough, with rare exceptions. We summarize the correction procedure as follows.

2.1. Single-matrix case

1. Solve the Sylvester equation (2.1.12) for X.

2. Perform the QR decomposition (2.1.13) to obtain unitary Q.

3. Execute the similarity transformation (2.1.14).

4. Check $\|\widehat{T}_{21}\|$ to determine whether another correction is needed.

The correction step looks a lot like the original swapping step. In fact, if the swapping step is initiated with the simple swap that creates (2.1.3), then the rest of the swapping step is identical to the correction step.

The special case $k = m = 1$

We now consider the most important special case, in which the blocks are 1×1:

$$T = \begin{bmatrix} t_{11} & t_{12} \\ & t_{22} \end{bmatrix},$$

and our task is to swap the eigenvalues $\lambda_1 = t_{11}$ and $\lambda_2 = t_{22}$. It turns out that this is always possible (even if $\lambda_1 = \lambda_2$), always stable, and refinement steps are never needed.

In this case equation (2.1.5) takes the form

$$\begin{bmatrix} t_{11} & t_{12} \\ & t_{22} \end{bmatrix} \begin{bmatrix} x \\ 1 \end{bmatrix} = \begin{bmatrix} x \\ 1 \end{bmatrix} t_{22},$$

and the Sylvester equation (2.1.6) becomes

$$x t_{22} - t_{11} x = t_{12},$$

a single linear equation in one unknown, whose solution is obviously

$$x = t_{12}/(t_{22} - t_{11}).$$

This gives us an eigenvector $\begin{bmatrix} x \\ 1 \end{bmatrix}$ associated with the eigenvalue t_{22}, but notice that any multiple of this vector is just as good, so we might as well skip the division by $t_{22} - t_{11}$ and use

$$\begin{bmatrix} x_1 \\ x_2 \end{bmatrix} = \begin{bmatrix} t_{12} \\ t_{22} - t_{11} \end{bmatrix}$$

instead. The QR decomposition (2.1.7) becomes

$$\begin{bmatrix} x_1 \\ x_2 \end{bmatrix} = \begin{bmatrix} c & -\overline{s} \\ s & \overline{c} \end{bmatrix} \begin{bmatrix} r \\ 0 \end{bmatrix}, \quad \text{where } |c|^2 + |s|^2 = 1. \quad (2.1.15)$$

All we need to do is compute c and s, and this can be done as follows:

$$\begin{array}{l} \nu \leftarrow \max\{|x_1|, |x_2|\} \\ \text{if } \nu = 0 \\ \quad \begin{bmatrix} c \leftarrow 1, \ s \leftarrow 0, \ r \leftarrow 0 \end{bmatrix} \\ \text{else} \\ \quad \begin{bmatrix} c \leftarrow x_1/\nu, \ s \leftarrow x_2/\nu \\ r \leftarrow \sqrt{|c|^2 + |s|^2} \\ c \leftarrow c/r, \ s \leftarrow s/r \\ r \leftarrow r\nu \end{bmatrix} \end{array} \quad (2.1.16)$$

This includes a small precaution to prevent overflow in the case of extremely large inputs and harmful underflow in the case of extremely small inputs.

It is easy to show that in floating-point arithmetic (2.1.16) computes c and s to high relative accuracy, assuming the input data x_1 and x_2 are accurate. Indeed the multiplications, divisions, and the single square root are all done with high relative accuracy. The only questionable step is the single addition, which could result in a high relative error if cancellation occurs. But cancellation cannot happen here because the numbers being added are both real and nonnegative. As a consequence, the computed quantities will be $\mathrm{fl}(c) = c(1+\varepsilon_1)$ and $\mathrm{fl}(s) = s(1+\varepsilon_2)$, where $|\varepsilon_1|$ and $|\varepsilon_2|$ are at most small multiples of the unit roundoff u.[4] The unitary similarity transformation (2.1.8) now takes the form

$$Q^*TQ = \begin{bmatrix} \bar{c} & \bar{s} \\ -s & c \end{bmatrix} \begin{bmatrix} t_{11} & t_{12} \\ & t_{22} \end{bmatrix} \begin{bmatrix} c & -\bar{s} \\ s & \bar{c} \end{bmatrix} = \begin{bmatrix} \hat{t}_{11} & \hat{t}_{12} \\ & \hat{t}_{22} \end{bmatrix},$$

with $\hat{t}_{11} = \lambda_2$ and $\hat{t}_{11} = \lambda_1$.

This transformation is backward stable, as unitary similarity transformations always are. We just have to check whether the $(2,1)$ entry of the result, which is zero in exact arithmetic, is small enough in floating-point arithmetic that we can set it to zero without compromising stability. In exact arithmetic we have

$$\hat{t}_{21} = -st_{11}c + (-st_{12} + ct_{22})s = ([t_{22} - t_{11}]c - t_{12}s)s = 0 \qquad (2.1.17)$$

because the vector $[c\ s]$ is proportional to $[t_{12}\ (t_{22} - t_{11})]$ by (2.1.15).

Plugging the computed values of c and s into (2.1.17) we get, in the first term, for example,

$$\mathrm{fl}(-st_{11}c) = -s(1+\varepsilon_2)t_{11}(1+\varepsilon_3)c(1+\varepsilon_1)(1+\varepsilon_4)(1+\varepsilon_5), \qquad (2.1.18)$$

where ε_3 and ε_4 are the roundoff errors in the first and second multiplications, respectively, and ε_5 is a roundoff error from the addition of this term to the second term. Each of these is from one floating-point operation and therefore satisfies $|\varepsilon_i| \leq u$, $i = 3, 4, 5$. Defining δ_1 by the equation

$$1 + \delta_1 = (1+\varepsilon_2)(1+\varepsilon_3)(1+\varepsilon_1)(1+\varepsilon_4)(1+\varepsilon_5),$$

we have $\delta_1 \approx \varepsilon_2 + \varepsilon_3 + \varepsilon_1 + \varepsilon_4 + \varepsilon_5$, so $|\delta_1| \leq Cu$, where C denotes a constant of modest size. We will use the notation $|\delta_1| \lesssim u$, which deemphasizes the constant, as an abbreviation. Now (2.1.18) takes the much simpler form $\mathrm{fl}(-st_{11}c) = -st_{11}c(1+\delta_1)$. Applying this treatment to the other terms in (2.1.17), we obtain

$$\mathrm{fl}(\hat{t}_{21}) = -st_{11}c(1+\delta_1) - st_{12}s(1+\delta_2) + ct_{22}s(1+\delta_3)$$
$$= 0 - st_{11}c\delta_1 - st_{12}s\delta_2 + ct_{22}s\delta_3,$$

where $|\delta_i| \lesssim u$, $i = 1, 2, 3$. Thus

$$|\mathrm{fl}(\hat{t}_{21})| \lesssim u\max\{|t_{11}|,|t_{12}|,|t_{22}|\} \approx u\|T\|,$$

suggesting that the entry can be safely set to zero. Years of numerical experience support this conclusion: The swap always succeeds, and a refinement step is never necessary. The reader can check that in the case $\lambda_1 = \lambda_2$, the transforming matrix Q is a diagonal matrix, which yields a trivial and harmless transformation.

---

[4] We use the standard model of floating-point arithmetic: If $*$ denotes any one of the four arithmetic operations, a and b are any two floating-point numbers, and we compute the quantity $a * b$, the computed quantity is denoted $\mathrm{fl}(a*b)$ and satisfies $\mathrm{fl}(a*b) = (a*b)(1+\varepsilon)$ with $|\varepsilon| \leq u$. In IEEE binary64 arithmetic, $u = 2^{-53} \approx 10^{-16}$. A good reference for roundoff errors and stability is Higham's book [35].

Other small cases

Since complex conjugate pairs of eigenvalues of a real matrix are often packaged as a single real 2×2 block, the 2×2 case is also quite important. Thus the problems of swapping a 2×2 block with either a 1×1 block (an eigenvalue) or another 2×2 block have been studied. Some relevant papers are [31, 51, 61, 64, 9, 10, 23].

The case $k = 2$, $m = 1$

Let us consider the case where

$$T = \begin{bmatrix} t_{11} & t_{12} & t_{13} \\ t_{21} & t_{22} & t_{23} \\ & & t_{33} \end{bmatrix},$$

and we want to find a unitary Q such that

$$Q^*TQ = \widehat{T} = \begin{bmatrix} \hat{t}_{11} & \hat{t}_{12} & \hat{t}_{13} \\ & \hat{t}_{22} & \hat{t}_{23} \\ & \hat{t}_{32} & \hat{t}_{33} \end{bmatrix},$$

with $\hat{t}_{11} = t_{33}$.

In the case $k = m = 1$, we solved the (trivial) Sylvester equation to get an eigenvector associated with the eigenvalue that we wanted to move upward. We then did a QR decomposition to get a unitary Q whose first column is proportional to that eigenvector. This Q is the desired transforming matrix. We can do the same thing in this case, but instead we will bypass the Sylvester equation and get right to the transforming matrix. What's needed is a Q such that

$$(T - t_{33}I)Q = \begin{bmatrix} 0 & * & * \\ 0 & * & * \\ 0 & * & * \end{bmatrix},$$

i.e., $(T - t_{33}I)Qe_1 = 0$, as then the first column of Q is an eigenvector associated with t_{33}. Noting that

$$T - t_{33}I = \begin{bmatrix} * & * & * \\ * & * & * \\ 0 & 0 & 0 \end{bmatrix},$$

we see that we can build Q in the form $Q_1 Q_2 \check{Q}_1$, where Q_1 is a core transformation that acts on columns 1 and 2, setting the $(2,1)$ entry to zero, Q_2 acts on columns 2 and 3, setting the $(2,2)$ entry to zero, and $\check{Q}_1$ sets the $(1,1)$ entry to zero. Each of these cores can be constructed using a variant of (2.1.16) (see Exercise 3).

We have $(T - t_{33}I)Qe_1 = 0$, so also $Q^*(T - t_{33}I)Qe_1 = 0$, and consequently

$$\widehat{T} = Q^*TQ = \begin{bmatrix} t_{33} & * & * \\ 0 & * & * \\ 0 & * & * \end{bmatrix},$$

as desired.

In floating-point arithmetic the $(2,1)$ and $(3,1)$ entries of $\widehat{T}$ will be slightly nonzero due to roundoff errors. As we did in the case $k = m = 1$, one can do a simple analysis to show that these entries are small enough that they can safely be set to zero without compromising backward stability. We leave the details to the reader.

The case $k = 1$, $m = 2$

Clearly the case $k = 1$, $m = 2$ is no more difficult than the case we just discussed, and an analogous procedure can be devised. See Exercise 4.

The case $k = 2$, $m = 2$

An early suggestion of Stewart [61] was to solve this case by using Francis's algorithm with exact shifts, but as of this writing the preferred method is to solve a Sylvester equation as described for general k and m above [9, 10, 23]. Refinement steps are sometimes needed.

Camps et al. [23] recently proposed a variant that uses a more elaborate computation in place of the QR decomposition. They showed this sometimes gives more accurate results.

Exercises 2.1

1. Given an $n \times n$ matrix A and an $m \times m$ matrix B, define the *Kronecker product* or *tensor product* of A and B to be the $nm \times nm$ matrix

$$A \otimes B = \begin{bmatrix} a_{11}B & \cdots & a_{1n}B \\ \vdots & & \vdots \\ a_{n1}B & \cdots & a_{nn}B \end{bmatrix}.$$

   (a) Suppose A and C are $n \times n$ and B and D are $m \times m$. Show that

   $$(A \otimes B)(C \otimes D) = AC \otimes BD.$$

   (b) Generalize the definition of tensor product to rectangular matrices, then state and prove a generalization of part (a).

   (c) Suppose A and B are nonsingular. Show that

   $$(A \otimes B)^{-1} = A^{-1} \otimes B^{-1}.$$

   (d) Show that if A is similar to S and B is similar to T, then $A \otimes B$ is similar to $S \otimes T$. Show further that if S and T are upper triangular, then $S \otimes T$ is upper triangular.

   (e) Prove that if A has eigenvalues $\lambda_1, \ldots, \lambda_n$ and B has eigenvalues $\mu_1, \ldots, \mu_m$, then $A \otimes B$ has eigenvalues $\lambda_i \mu_j$, $i = 1, \ldots, n$, $j = 1, \ldots, m$.

   (f) Prove that $A \otimes I_m - I_n \otimes B$ has eigenvalues $\lambda_i - \mu_j$, $i = 1, \ldots, n$, $j = 1, \ldots, m$. Therefore $A \otimes I_m - I_n \otimes B$ is nonsingular if and only if A and B have no common eigenvalues.

2. The previous exercise is a prerequisite for this one, in which we prove that the Sylvester equation (2.1.6), i.e., $XT_{22} - T_{11}X = T_{12}$, has a unique solution if the spectra of T_{11} and T_{22} are disjoint. If X is $k \times m$ with columns $x_1, \ldots, x_m$, we define the vector $\text{vec}(X)$ in $\mathbb{C}^{km}$ by

$$\text{vec}(X) = \begin{bmatrix} x_1 \\ \vdots \\ x_m \end{bmatrix}.$$

   (a) Prove that (2.1.6) can be written as $Mw = b$, where

   $$M = T_{22}^T \otimes I_k - I_m \otimes T_{11},$$

   $w = \text{vec}(X)$, and $b = \text{vec}(T_{12})$.

(b) Deduce that (2.1.6) has a unique solution if and only if T_{11} and T_{22} have no common eigenvalues.

3. Write an algorithm like (2.1.16) that computes c and s with $|c|^2 + |s|^2 = 1$ such that

$$\begin{bmatrix} x_1 & x_2 \end{bmatrix} \begin{bmatrix} c & -\overline{s} \\ s & \overline{c} \end{bmatrix} = \begin{bmatrix} 0 & r \end{bmatrix},$$

where $\begin{bmatrix} x_1 & x_2 \end{bmatrix}$ is given and $r = \|x\|$.

4. Consider the case $k = 1$, $m = 2$. We have

$$T = \begin{bmatrix} t_{11} & t_{12} & t_{13} \\ & t_{22} & t_{23} \\ & t_{32} & t_{33} \end{bmatrix},$$

and we wish to move the eigenvalue t_{11} to the bottom right corner. Show how to construct a unitary Q, a product of three core transformations, such that $e_3^T Q^*(T - t_{11}I) = 0$. (The third row of Q^* is a left eigenvector of T associated with t_{11}.) Show that

$$Q^*TQ = \begin{bmatrix} * & * & * \\ * & * & * \\ 0 & 0 & t_{11} \end{bmatrix},$$

as desired.

5. In this exercise we discuss an alternative to swapping blocks when computing the eigenvectors of a matrix in Schur form T.

    Let $\lambda_i = t_{ii}$ be an eigenvalue of T. For simplicity assume that all eigenvalues are pairwise disjoint. The corresponding eigenvector v_i is a nonzero vector in the null space of $T - \lambda_i I$.

    (a) Show that $e_j^T v_i = 0$ for all $j > i$.
    (b) The vector v can be scaled and we choose $e_i^T v_i = 1$. Explain how to compute the remaining entries of v_i.
    (c) Why is $e_i^T v_i = 1$ a reasonable assumption?

2.2 ▪ Pencil case

Let us now consider the problem of swapping eigenvalues in a triangular pencil or, more generally, blocks in a block-triangular pencil. We will start with the general case of blocks of arbitrary size. Consider therefore a block-triangular pencil

$$C - \lambda F = \begin{bmatrix} C_{11} & C_{12} \\ & C_{22} \end{bmatrix} - \lambda \begin{bmatrix} F_{11} & F_{12} \\ & F_{22} \end{bmatrix}$$

with $C_{11}, F_{11} \in \mathbb{C}^{k \times k}$ and $C_{22}, F_{22} \in \mathbb{C}^{m \times m}$. We will assume that the subpencils $C_{11} - \lambda F_{11}$ and $C_{22} - \lambda F_{22}$ have disjoint spectra, i.e.,

$$\Lambda(C_{11}, F_{11}) \cap \Lambda(C_{22}, F_{22}) = \emptyset.$$

We seek a unitary equivalence transformation

$$Q^*(C - \lambda F)Z = \widehat{C} - \lambda \widehat{F} = \begin{bmatrix} \widehat{C}_{11} & \widehat{C}_{12} \\ & \widehat{C}_{22} \end{bmatrix} - \lambda \begin{bmatrix} \widehat{F}_{11} & \widehat{F}_{12} \\ & \widehat{F}_{22} \end{bmatrix}$$

with $\Lambda(\widehat{C}_{11}, \widehat{F}_{11}) = \Lambda(C_{22}, F_{22})$ and $\Lambda(\widehat{C}_{22}, \widehat{F}_{22}) = \Lambda(C_{11}, F_{11})$.

We proceed as in the single-matrix case. Applying a similarity transformation (a special equivalence) by
$$\begin{bmatrix} 0 & I_k \\ I_m & 0 \end{bmatrix}$$
to the pencil, we reverse the blocks to get
$$\begin{bmatrix} C_{22} & \\ C_{12} & C_{11} \end{bmatrix} - \lambda \begin{bmatrix} F_{22} & \\ F_{12} & F_{11} \end{bmatrix}.$$

This is the desired swap, except that the blocks C_{12} and F_{12} are out of place. We now seek a simple equivalence transformation that gets rid of these blocks; specifically, we seek matrices X, $Y \in \mathbb{C}^{k \times m}$ such that
$$\begin{bmatrix} I_m & \\ Y & I_k \end{bmatrix}^{-1} \begin{bmatrix} C_{22} & \\ C_{12} & C_{11} \end{bmatrix} \begin{bmatrix} I_m & \\ X & I_k \end{bmatrix} = \begin{bmatrix} C_{22} & \\ & C_{11} \end{bmatrix}$$
and
$$\begin{bmatrix} I_m & \\ Y & I_k \end{bmatrix}^{-1} \begin{bmatrix} F_{22} & \\ F_{12} & F_{11} \end{bmatrix} \begin{bmatrix} I_m & \\ X & I_k \end{bmatrix} = \begin{bmatrix} F_{22} & \\ & F_{11} \end{bmatrix}.$$

After some simple algebra we find that these equations are equivalent to the *generalized Sylvester equations*
$$YC_{22} - C_{11}X = C_{12}, \qquad YF_{22} - F_{11}X = F_{12}. \tag{2.2.1}$$

This is a system of $2mk$ linear equations in $2mk$ unknowns, namely the entries of X and Y. The following theorem is proved in Exercise 1.

Theorem 2.2.1. *The generalized Sylvester equations (2.2.1) have a unique solution if and only if the spectra of the pairs (C_{11}, F_{11}) and (C_{22}, F_{22}) are disjoint.*

In terms of the original (unflipped) coordinate system, the equivalence transformation can be written as
$$\begin{bmatrix} C_{11} & C_{12} \\ & C_{22} \end{bmatrix} \begin{bmatrix} I_k & X \\ & I_m \end{bmatrix} = \begin{bmatrix} I_k & Y \\ & I_m \end{bmatrix} \begin{bmatrix} C_{11} & \\ & C_{22} \end{bmatrix},$$
and similarly for F. The nontrivial part of this equation is the second block column, which is
$$\begin{bmatrix} C_{11} & C_{12} \\ & C_{22} \end{bmatrix} \begin{bmatrix} X \\ I_m \end{bmatrix} = \begin{bmatrix} Y \\ I_m \end{bmatrix} C_{22},$$
and similarly
$$\begin{bmatrix} F_{11} & F_{12} \\ & F_{22} \end{bmatrix} \begin{bmatrix} X \\ I_m \end{bmatrix} = \begin{bmatrix} Y \\ I_m \end{bmatrix} F_{22}.$$

This shows (by Proposition 1.2.5) that when we solve the generalized Sylvester equations (2.2.1) for X and Y, we are computing a deflating pair
$$\left(\begin{bmatrix} X \\ I_m \end{bmatrix}, \begin{bmatrix} Y \\ I_m \end{bmatrix} \right)$$
for (C, F) associated with $\Lambda(C_{22}, F_{22})$. Combining the transformation matrices to form
$$\begin{bmatrix} & I_k \\ I_m & \end{bmatrix} \begin{bmatrix} I_m & \\ X & I_k \end{bmatrix} = \begin{bmatrix} X & I_k \\ I_m & \end{bmatrix},$$

2.2. Pencil case

and similarly for Y, we have

$$\begin{bmatrix} Y & I_k \\ I_m & \end{bmatrix}^{-1} \begin{bmatrix} C_{11} & C_{12} \\ & C_{22} \end{bmatrix} \begin{bmatrix} X & I_k \\ I_m & \end{bmatrix} = \begin{bmatrix} C_{22} & \\ & C_{11} \end{bmatrix},$$

and similarly for F. This solves the block-exchange problem, except that the transforming matrices are not unitary. To get a unitary equivalence, we take two QR decompositions,

$$\begin{bmatrix} X \\ I_m \end{bmatrix} = ZS = \begin{bmatrix} Z_1 & Z_2 \end{bmatrix} \begin{bmatrix} S_{11} \\ 0 \end{bmatrix} = Z_1 S_{11} \qquad (2.2.2)$$

and

$$\begin{bmatrix} Y \\ I_m \end{bmatrix} = QR = \begin{bmatrix} Q_1 & Q_2 \end{bmatrix} \begin{bmatrix} R_{11} \\ 0 \end{bmatrix} = Q_1 R_{11}, \qquad (2.2.3)$$

and use the unitary Z and Q as our transforming matrices. Since the columns of Z_1 (resp., Q_1) span the same space as the columns of $\begin{bmatrix} X \\ I_m \end{bmatrix}$ (resp., $\begin{bmatrix} Y \\ I_m \end{bmatrix}$), we deduce that (Z_1, Q_1) is a deflating pair (associated with $\Lambda(C_{22}, F_{22})$) for the pencil $C - \lambda F$, and therefore

$$Q^*(C - \lambda F)Z = \begin{bmatrix} \widehat{C}_{11} & \widehat{C}_{12} \\ & \widehat{C}_{22} \end{bmatrix} - \lambda \begin{bmatrix} \widehat{F}_{11} & \widehat{F}_{12} \\ & \widehat{F}_{22} \end{bmatrix} \qquad (2.2.4)$$

with $\Lambda(\widehat{C}_{11}, \widehat{F}_{11}) = \Lambda(C_{22}, F_{22})$. This completes the description of the exchange, which can be summarized as follows.

1. Compute X and Y by solving the generalized Sylvester equations (2.2.1).

2. Compute the QR decompositions (2.2.2) and (2.2.3).

3. Execute the equivalence (2.2.4).

Refinement

When we compute the equivalence (2.2.4) in floating-point arithmetic, the result will not be exactly block triangular due to roundoff errors. Instead we get

$$\begin{bmatrix} \widehat{C}_{11} & \widehat{C}_{12} \\ \widehat{C}_{21} & \widehat{C}_{22} \end{bmatrix} - \lambda \begin{bmatrix} \widehat{F}_{11} & \widehat{F}_{12} \\ \widehat{F}_{21} & \widehat{F}_{22} \end{bmatrix},$$

where $\|\widehat{C}_{21}\|$ and $\|\widehat{F}_{21}\|$ are tiny but nonzero. If they are small enough, we can simply set $\widehat{C}_{21}$ and $\widehat{F}_{21}$ to zero and move on.

If they are not small enough, a refinement procedure can be used to drive them to zero. We now describe that procedure. For simplicity we drop the hats from the symbols and consider a nearly block-triangular pencil

$$\begin{bmatrix} C_{11} & C_{12} \\ C_{21} & C_{22} \end{bmatrix} - \lambda \begin{bmatrix} F_{11} & F_{12} \\ F_{21} & F_{22} \end{bmatrix},$$

with $\|C_{21}\|$ and $\|F_{21}\|$ tiny relative to $\|C\|$ and $\|F\|$, respectively. Our objective is to construct an equivalence that changes the pencil only slightly, transforming C_{21} and F_{21} to zero. Based on our experience so far, we seek $X, Y \in \mathbb{C}^{k \times m}$, both with tiny norm, such that

$$\begin{bmatrix} I_m & \\ Y & I_k \end{bmatrix}^{-1} \begin{bmatrix} C_{11} & C_{12} \\ C_{21} & C_{22} \end{bmatrix} \begin{bmatrix} I_m & \\ X & I_k \end{bmatrix} = \begin{bmatrix} \widetilde{C}_{11} & \widetilde{C}_{12} \\ & \widetilde{C}_{22} \end{bmatrix}$$

and
$$\begin{bmatrix} I_m & \\ Y & I_k \end{bmatrix}^{-1} \begin{bmatrix} F_{11} & F_{12} \\ F_{21} & F_{22} \end{bmatrix} \begin{bmatrix} I_m & \\ X & I_k \end{bmatrix} = \begin{bmatrix} \widetilde{F}_{11} & \widetilde{F}_{12} \\ & \widetilde{F}_{22} \end{bmatrix}.$$

Minor manipulations show that these equations are equivalent to
$$\begin{bmatrix} C_{11} & C_{12} \\ C_{21} & C_{22} \end{bmatrix} \begin{bmatrix} I_m \\ X \end{bmatrix} = \begin{bmatrix} I_m \\ Y \end{bmatrix} \widetilde{C}_{11}$$

and
$$\begin{bmatrix} F_{11} & F_{12} \\ F_{21} & F_{22} \end{bmatrix} \begin{bmatrix} I_m \\ X \end{bmatrix} = \begin{bmatrix} I_m \\ Y \end{bmatrix} \widetilde{F}_{11},$$

which means that we are seeking a deflating pair
$$\left(\begin{bmatrix} I_m \\ X \end{bmatrix}, \begin{bmatrix} I_m \\ Y \end{bmatrix} \right)$$

associated with the spectrum of $(\widetilde{C}_{11}, \widetilde{F}_{11})$, a slight perturbation of (C_{11}, F_{11}). Additional minor manipulations show that these equations are in turn equivalent to a pair of *generalized algebraic Riccati equations*

$$YC_{11} - C_{22}X = C_{21} - YC_{12}X, \qquad YF_{11} - F_{22}X = F_{21} - YF_{12}X. \tag{2.2.5}$$

This quadratic system of equations may have many solutions, but one can show, as in the single-matrix case, that there is exactly one solution for which $\|X\|$ and $\|Y\|$ are tiny, provided $\|C_{21}\|$ and $\|F_{21}\|$ are sufficiently small. This is the solution that is of interest to us.

For small enough X and Y, the quadratic terms $YC_{12}X$ and $YF_{12}X$ in (2.2.5) will be negligible in size. Therefore we can approximate X and Y well by dropping the quadratic terms and solving the resulting generalized Sylvester equations

$$YC_{11} - C_{22}X = C_{21}, \qquad YF_{11} - F_{22}X = F_{21}, \tag{2.2.6}$$

which are of exactly the same form (after adjustment of subscripts) as (2.2.1). Once we have solved these equations, the resulting X and Y are not exactly the ones we want, but they will be very close. This is one step of Newton's method applied to the quadratic equation (2.2.5) with initial guesses $X^{(0)} = Y^{(0)} = 0$.

Once we have X and Y, we proceed as in the original swapping step and compute QR decompositions
$$\begin{bmatrix} I_m \\ X \end{bmatrix} = ZS = \begin{bmatrix} Z_1 & Z_2 \end{bmatrix} \begin{bmatrix} S_{11} \\ 0 \end{bmatrix} = Z_1 S_{11} \tag{2.2.7}$$

and
$$\begin{bmatrix} I_m \\ Y \end{bmatrix} = QR = \begin{bmatrix} Q_1 & Q_2 \end{bmatrix} \begin{bmatrix} R_{11} \\ 0 \end{bmatrix} = Q_1 R_{11}, \tag{2.2.8}$$

then make the unitary equivalence
$$Q^*(C - \lambda F)Z = \begin{bmatrix} \widetilde{C}_{11} & \widetilde{C}_{12} \\ \widetilde{C}_{21} & \widetilde{C}_{22} \end{bmatrix} - \lambda \begin{bmatrix} \widetilde{F}_{11} & \widetilde{F}_{12} \\ \widetilde{F}_{21} & \widetilde{F}_{22} \end{bmatrix}. \tag{2.2.9}$$

If we had obtained X and Y by solving (2.2.5) exactly, and if we had made no roundoff errors anywhere in the computation, the pair $(\begin{bmatrix} I_m \\ X \end{bmatrix}, \begin{bmatrix} I_m \\ Y \end{bmatrix})$ would be exactly a deflating pair for (C, F), and so would (Z_1, Q_1). Therefore the equivalence transformation would yield $\widetilde{C}_{21} = \widetilde{F}_{21} = 0$.

2.2. Pencil case

Since we do not solve (2.2.5) exactly, and we do make further roundoff errors, $\widetilde{C}_{21}$ and $\widetilde{F}_{21}$ will not be exactly zero. If $\|\widetilde{C}_{21}\|$ and $\|\widetilde{F}_{21}\|$ are small enough, we can set them to zero and move on. If not, we can repeat the refinement step. Since Newton's method converges quadratically, very few corrections are necessary. In fact, one correction will almost always be enough. We summarize the refinement procedure as follows.

1. Solve the generalized Sylvester equations (2.2.6) for X and Y.

2. Perform the QR decompositions (2.2.7) and (2.2.8) to obtain unitary Z and Q.

3. Execute the equivalence transformation (2.2.9).

4. Check $\|\widetilde{C}_{21}\|$ and $\|\widetilde{F}_{21}\|$ to determine whether another correction is needed.

The case $m = k = 1$

Once again, the most important special case is $m = k = 1$. Consider a 2×2 pencil

$$C - \lambda F = \begin{bmatrix} c_{11} & c_{12} \\ & c_{22} \end{bmatrix} - \lambda \begin{bmatrix} f_{11} & f_{12} \\ & f_{22} \end{bmatrix} \qquad (2.2.10)$$

with eigenvalues $\lambda_1 = c_{11}/f_{11}$ and $\lambda_2 = c_{22}/f_{22}$. We wish to transform this by a unitary equivalence to a pencil

$$\widehat{C} - \lambda \widehat{F} = \begin{bmatrix} \hat{c}_{11} & \hat{c}_{12} \\ & \hat{c}_{22} \end{bmatrix} - \lambda \begin{bmatrix} \hat{f}_{11} & \hat{f}_{12} \\ & \hat{f}_{22} \end{bmatrix} \qquad (2.2.11)$$

with $\hat{c}_{11}/\hat{f}_{11} = \lambda_2$ and $\hat{c}_{22}/\hat{f}_{22} = \lambda_1$.

If we proceed as described in the case of general m and k above, we must solve the generalized Sylvester equations (2.2.1), which reduce to a system of two equations in two unknowns

$$\begin{bmatrix} c_{11} & c_{22} \\ f_{11} & f_{22} \end{bmatrix} \begin{bmatrix} -x \\ y \end{bmatrix} = \begin{bmatrix} c_{12} \\ f_{12} \end{bmatrix}$$

in this case. If we solve this for x and y, then $([\begin{smallmatrix}x\\1\end{smallmatrix}], [\begin{smallmatrix}y\\1\end{smallmatrix}])$ is a deflating pair for $C - \lambda F$ associated with the eigenvalue λ_2. Our next task would be to compute two QR decompositions (2.2.2) and (2.2.3), but these are now very simple as the matrices are 2×2. Finally we would use the resulting Z and Q to execute the unitary equivalence (2.2.4). This works quite well, and it can be improved by refinement steps if necessary. The refinement step is equally simple in this case, but we will not describe it in detail.

Backward stable variant

We will describe instead a slightly different swapping procedure (valid for the special case $m = k = 1$) that works better in the real world of floating-point computation. This method was described in [21] and is a variant of a method due to Van Dooren [64]. It is backward stable, it never fails, even if the eigenvalues are equal, and it never requires a refinement step.

We start with a 2×2 pencil $C - \lambda F$ as in (2.2.10). We wish to transform it to (2.2.11), which has the eigenvalues reversed. We need to compute unitary Q and Z such that $\widehat{C} - \lambda \widehat{F} = Q^*(C - \lambda F)Z$. We will present two methods, which are duals of one another. In Chapter 5 on the standard eigenvalue problem we will use both of them.

Primal method (Z first)

Substituting λ_2 for λ in (2.2.10), we obtain

$$H = f_{22}C - c_{22}F = \begin{bmatrix} f_{22}c_{11} - c_{22}f_{11} & f_{22}c_{12} - c_{22}f_{12} \\ 0 & 0 \end{bmatrix} = \begin{bmatrix} * & * \\ 0 & 0 \end{bmatrix}.$$

Now compute a core transformation Z such that

$$HZ = \begin{bmatrix} 0 & * \\ 0 & 0 \end{bmatrix}.$$

This is done by a variant of (2.1.16). This Z satisfies

$$(f_{22}C - c_{22}F)Ze_1 = 0, \tag{2.2.12}$$

so $z_1 = Ze_1$ is a right eigenvector of (C, F) associated with the eigenvalue $\lambda_2 = c_{22}/f_{22}$. This is just what we need to bring λ_2 to the top. Equation (2.2.12) implies that Cz_1 and Fz_1 are proportional, so there is a core transformation Q such that both $Q^*Cz_1 = \alpha e_1$ and $Q^*Fz_1 = \beta e_1$. Letting $\widehat{C} - \lambda \widehat{F} = Q^*(C - \lambda F)Z$, we see that $\hat{c}_{21} = q_2^*Cz_1 = 0$ and $\hat{f}_{21} = q_2^*Fz_1 = 0$, so the transformed pencil is triangular, as desired. Also, from (2.2.12),

$$f_{22}\hat{c}_{11} - c_{22}\hat{f}_{11} = q_1^*(f_{22}C - c_{22}F)z_1 = 0,$$

showing that the eigenvalue $\lambda_2 = c_{22}/f_{22}$ has been moved to the top.

The stability of the algorithm depends upon whether Cz_1 or Fz_1 is used in the computation of Q. In Section 2.3 we will show that if $|\lambda_1| \geq |\lambda_2|$ and Fz_1 is used to compute Q, the swap is backward stable. But what if $|\lambda_1| < |\lambda_2|$? Then we can simply interchange the roles of F and C to invert the eigenvalues. All this means is that we use Cz_1 instead of Fz_1 to compute Q.

We summarize the procedure as follows:

- Compute the first row of $H = f_{22}C - c_{22}F = \begin{bmatrix} * & * \\ 0 & 0 \end{bmatrix}$.

- Compute core Z such that $HZ = \begin{bmatrix} 0 & * \\ 0 & 0 \end{bmatrix}$ by a variant of (2.1.16).

- Compute $y = \begin{cases} FZe_1 & \text{if } |c_{11}f_{22}| \geq |c_{22}f_{11}| \text{ (i.e., } |\lambda_1| \geq |\lambda_2|\text{)}, \\ CZe_1 & \text{otherwise.} \end{cases}$

- Compute core Q such that $Q^*y = \alpha e_1$ by (2.1.16).

- Execute the unitary equivalence $\widehat{C} - \lambda \widehat{F} = Q^*(C - \lambda F)Z$.

This algorithm never breaks down, even if $\lambda_1 = \lambda_2$. The reader can check that in that case we get a trivial transformation that leaves the eigenvalues where they are.

This is almost exactly the procedure proposed by Van Dooren [64]. The only difference is in the criterion for deciding when to switch between F and C for the computation of y. Van Dooren's procedure uses FZe_1 if $|f_{22}| \geq |c_{22}|$ and CZe_1 otherwise. The method works excellently in practice, and years of practical experience suggest that refinement steps are never needed. Van Dooren [64] proved that his procedure is backward stable in the sense that the backward errors are bounded by a modest multiple of $u \max\{\|C\|, \|F\|\}$.

We will show that our method is backward stable in a stronger sense: The backward errors in C (resp., F) are bounded by a modest multiple of $u\|C\|$ (resp., $u\|F\|$). This occasionally makes a difference. Some numerical experiments in [21] show that our method performs a bit better in extreme situations.

We defer the proof to Section 2.3

2.2. Pencil case

Dual method (Q first)

Substituting λ_1 for λ in (2.2.10), we obtain

$$H = f_{11}C - c_{11}F = \begin{bmatrix} 0 & f_{11}c_{12} - c_{11}f_{12} \\ 0 & f_{11}c_{22} - c_{11}f_{22} \end{bmatrix} = \begin{bmatrix} 0 & * \\ 0 & * \end{bmatrix}.$$

Now use (2.1.16) to compute a core transformation Q such that

$$Q^*H = \begin{bmatrix} 0 & * \\ 0 & 0 \end{bmatrix}.$$

This Q satisfies

$$e_2^* Q^* (f_{11}C - c_{11}F) = 0, \qquad (2.2.13)$$

so $q_2^* = e_2^* Q^*$ is a left eigenvector of (C, F) associated with the eigenvalue $\lambda_1 = c_{11}/f_{11}$. This is just what we need to bring λ_1 to the bottom. Equation (2.2.13) implies that $q_2^* C$ and $q_2^* F$ are proportional, so there is a core transformation Z such that both $q_2^* C Z = \alpha e_2^*$ and $q_2^* F Z = \beta e_2^*$. Letting $\widehat{C} - \lambda \widehat{F} = Q^*(C - \lambda F)Z$, we see that $\hat{c}_{21} = q_2^* C z_1 = 0$ and $\hat{f}_{21} = q_2^* F z_1 = 0$, so the transformed pencil is triangular, as desired. Also, from (2.2.13),

$$f_{11} \hat{c}_{22} - c_{11} \hat{f}_{22} = q_2^* (f_{11}C - c_{11}F) z_2 = 0,$$

showing that the eigenvalue $\lambda_1 = c_{11}/f_{11}$ has been moved to the bottom.

It is not hard to show (Exercise 2) that the dual algorithm is equivalent to the primal algorithm applied to the pertransposed pencil

$$\begin{bmatrix} \bar{c}_{22} & \bar{c}_{12} \\ 0 & \bar{c}_{11} \end{bmatrix} - \lambda \begin{bmatrix} \bar{f}_{22} & \bar{f}_{12} \\ 0 & \bar{f}_{11} \end{bmatrix},$$

so stability of the primal algorithm implies stability of the dual algorithm. One must take into account that the roles of λ_1 and λ_2 are reversed, as are those of Q and Z. For stability, $q_2^* F$ should be used to determine Z if $|\lambda_2| > |\lambda_1|$. Otherwise $q_2^* C$ should be used.

We summarize the procedure as follows:

- Compute the second column of $H = f_{11}C - c_{11}F = \begin{bmatrix} 0 & * \\ 0 & * \end{bmatrix}$.

- Compute core Q such that $Q^*H = \begin{bmatrix} 0 & * \\ 0 & 0 \end{bmatrix}$ by (2.1.16).

- Compute $y^* = \begin{cases} q_2^* C & \text{if } |c_{11} f_{22}| \geq |c_{22} f_{11}| \text{ (i.e. } |\lambda_1| \geq |\lambda_2|\text{)}, \\ q_2^* F & \text{otherwise.} \end{cases}$

- Compute core Z such that $y^* Z = \alpha e_2^*$ by a variant of (2.1.16).

- Execute the unitary equivalence $\widehat{C} - \lambda \widehat{F} = Q^*(C - \lambda F)Z$.

The case $k = 2$, $m = 1$

Swapping 2×2 blocks is also of great interest, especially if C and F are real. Let's take a look at the case $k = 2$, $m = 1$. (Clearly, if we can do this case, we will also be able to do the case $k = 1$, $m = 2$ by an analogous procedure.) We have

$$C - \lambda F = \begin{bmatrix} c_{11} & c_{12} & c_{13} \\ c_{21} & c_{22} & c_{23} \\ & & c_{33} \end{bmatrix} - \lambda \begin{bmatrix} f_{11} & f_{12} & f_{13} \\ f_{21} & f_{22} & f_{23} \\ & & f_{33} \end{bmatrix},$$

which we want to transform to

$$\widehat{C} - \lambda \widehat{F} = \begin{bmatrix} \hat{c}_{11} & \hat{c}_{12} & \hat{c}_{13} \\ & \hat{c}_{22} & \hat{c}_{23} \\ & \hat{c}_{32} & \hat{c}_{33} \end{bmatrix} - \lambda \begin{bmatrix} \hat{f}_{11} & \hat{f}_{12} & \hat{f}_{13} \\ & \hat{f}_{22} & \hat{f}_{23} \\ & \hat{f}_{32} & \hat{f}_{33} \end{bmatrix},$$

with $\hat{c}_{11}/\hat{f}_{11} = \lambda_3 = c_{33}/f_{33}$. Of course, we can use the general procedures outlined at the beginning of the section, but there is also a method that is close to the one we just described for the case $k = m = 1$. Let

$$H = f_{33}C - c_{33}F = \begin{bmatrix} h_{11} & h_{12} & h_{13} \\ h_{21} & h_{22} & h_{23} \\ 0 & 0 & 0 \end{bmatrix}.$$

If we solve the underdetermined system $Hx = 0$, we get an eigenvector associated with the eigenvalue $\lambda_3 = c_{33}/f_{33}$, which we can use to determine the transforming matrices that we need. To this end we will build a unitary Z such that the first column of HZ is zero, i.e., $HZe_1 = 0$. Then Ze_1 is an eigenvector of (C, F) associated with eigenvalue λ_3.

We begin by determining a core $\check{Q}_1$ that acts on rows 1 and 2 to annihilate h_{21}:

$$\check{Q}_1^* H = \begin{bmatrix} d_1 & * & * \\ 0 & d_2 & * \\ 0 & 0 & 0 \end{bmatrix}. \tag{2.2.14}$$

Then we determine Z_2 that acts on columns 2 and 3 to annihilate d_2, and Z_1 that acts on columns 1 and 2 to annihilate d_1. Let $Z = Z_2 Z_1$. Then the first column of $\check{Q}_1^* HZ$ is zero, i.e., $\check{Q}_1^* HZe_1 = 0$, and therefore also $HZe_1 = 0$.

Because Ze_1 is an eigenvector of (C, F), the vectors CZe_1 and FZe_1 must be proportional in exact arithmetic. If $\sqrt{|\lambda_1 \lambda_2|} \geq |\lambda_3|$,[5] compute $y = FZe_1$. Otherwise, compute $y = CZe_1$. In other words, reverse the roles of F and C.

Next compute Q_2 such that $Q_2^* y$ has a zero in the third position and Q_1 so that $Q_1^* Q_2^* y$ has a zero in the second position. In fact this is just a QR decomposition $y = Q \begin{bmatrix} r \\ 0 \\ 0 \end{bmatrix}$, where $Q = Q_2 Q_1$. Finally compute

$$\widehat{C} = Q^* CZ = Q_1^* Q_2^* C Z_2 Z_1 = \begin{bmatrix} \hat{c}_{11} & \hat{c}_{12} & \hat{c}_{13} \\ & \hat{c}_{22} & \hat{c}_{23} \\ & \hat{c}_{32} & \hat{c}_{33} \end{bmatrix} \tag{2.2.15}$$

and

$$\widehat{F} = Q^* FZ = Q_1^* Q_2^* F Z_2 Z_1 = \begin{bmatrix} \hat{f}_{11} & \hat{f}_{12} & \hat{f}_{13} \\ & \hat{f}_{22} & \hat{f}_{23} \\ & \hat{f}_{32} & \hat{f}_{33} \end{bmatrix}. \tag{2.2.16}$$

One easily checks that $\hat{c}_{11}/\hat{f}_{11} = \lambda_3$. The procedure is summarized as follows:

- Compute $H = f_{33}C - c_{33}F$.

- Compute $\check{Q}_1$, Z_2, and Z_1 as described in and around (2.2.14).

- Compute $y = \begin{cases} FZe_1 & \text{if } \sqrt{|\lambda_1 \lambda_2|} \geq |\lambda_3|, \\ CZe_1 & \text{otherwise.} \end{cases}$

---

[5]In the most common application, C and F are real, and λ_1 and λ_2 are a complex conjugate pair, so $|\lambda_1| = |\lambda_2| = \sqrt{|\lambda_1 \lambda_2|}$.

2.2. Pencil case

- Compute Q_2 and Q_1 via the decomposition $y = Q_2 Q_1 R = QR$.

- Execute the unitary equivalence (2.2.15) and (2.2.16).

Van Dooren [64] presented a very similar method. His criterion for using FZ rather than CZ for the computation of Q was $|f_{33}| \geq |c_{33}|$. He briefly sketched a proof of backward stability of the method, in the sense that the backward errors are $\lesssim u \max\{\|C\|, \|F\|\}$. The criterion $\sqrt{|\lambda_1 \lambda_2|} \geq |\lambda_3|$ that we have suggested is an heuristic based on its success in the case $k = m = 1$.

Experience suggests that refinement steps are never necessary.

The case $k = 2, m = 2$

In this case the best advice seems to be to use the general procedure described at the beginning of the section. Refinement steps are sometimes necessary. Some works that discuss this case are [64, 37, 38, 23].

Camps et al. [23] studied this case and suggested a variant that replaces the QR decomposition step by a more elaborate computation. This gives more accurate results in some extreme cases.

Exercises 2.2

1. In this exercise you will prove Theorem 2.2.1. Exercises 2.1.1 and 2.1.2 are prerequisites.

   (a) Apply the Generalized Schur Theorem (Theorem 1.2.1) to the pairs (C_{11}^T, F_{11}^T) and (C_{22}, F_{22}) to show that the generalized Sylvester equations (2.2.1) can be replaced by an equivalent system in which C_{11} and F_{11} are lower triangular and C_{22} and F_{22} are upper triangular.

   (b) Using properties of Kronecker products, show that the generalized Sylvester equations (2.2.1) can be rewritten as

   $$\begin{bmatrix} I \otimes C_{11} & C_{22}^T \otimes I \\ I \otimes F_{11} & F_{22}^T \otimes I \end{bmatrix} \begin{bmatrix} -\text{vec}(X) \\ \text{vec}(Y) \end{bmatrix} = \begin{bmatrix} \text{vec}(C_{12}) \\ \text{vec}(F_{12}) \end{bmatrix}. \quad (2.2.17)$$

   (c) From part (a) we can assume that C_{11}, F_{11}, C_{22}^T, and F_{22}^T are all lower triangular in (2.2.17). Assume this for the remainder of the exercise. Show that each of the four blocks in the coefficient matrix is lower triangular.

   (d) Transform (2.2.17) to an equivalent system by doing a perfect shuffle of the rows and columns. This means that rows (and columns) that were ordered $1, 2, \ldots, 2n$ get reordered as $1, n+1, 2, n+2, \ldots, n, 2n$. Show that the resulting system is block lower triangular with 2×2 blocks. Show that each main-diagonal block has the form

   $$\begin{bmatrix} c_1 & c_2 \\ f_1 & f_2 \end{bmatrix},$$

   where c_i/f_i is an eigenvalue of (C_{ii}, F_{ii}) for $i = 1, 2$. Deduce that the coefficient matrix is nonsingular if and only if (C_{11}, F_{11}) and (C_{22}, F_{22}) have no eigenvalues in common. This completes the proof of Theorem 2.2.1.

2. Let $C - \lambda F = \begin{bmatrix} c_{11} & c_{12} \\ & c_{22} \end{bmatrix} - \lambda \begin{bmatrix} f_{11} & f_{12} \\ & f_{22} \end{bmatrix}$ and $P = \begin{bmatrix} 0 & 1 \\ 1 & 0 \end{bmatrix}$.

(a) Show that $PC^*P - \lambda PF^*P = \begin{bmatrix} \bar{c}_{22} & \bar{c}_{12} \\ & \bar{c}_{11} \end{bmatrix} - \lambda \begin{bmatrix} \bar{f}_{22} & \bar{f}_{12} \\ & \bar{f}_{11} \end{bmatrix}$.

This is the *pertransposed* pencil.

(b) If we now apply the primal method to swap the eigenvalues in this pertransposed pencil, we get $\widehat{C} - \lambda \widehat{F} = Q^*(PC^*P - \lambda PF^*P)Z$. Show that we transform back to the original coordinate system by taking $P\widehat{C}^*P - \lambda P\widehat{F}^*P$.

(c) Show that $P\widehat{C}^*P - \lambda P\widehat{F}^*P = (PZP)^*(C - \lambda F)(PQP)$. The roles of Q and Z have been reversed.

(d) Conclude that applying the primal method to the pertransposed pencil is equivalent to applying the dual method to the original pencil, as claimed in the text.

3. Devise a method for interchanging blocks in the case $k = 1$, $m = 2$. Start with $H = f_{11}C - c_{11}F$, which has all zeros in the first column. We want to transform the last row to zero.

2.3 ▪ Backward stability in the case $m = k = 1$

We have a pencil
$$C - \lambda F = \begin{bmatrix} c_{11} & c_{12} \\ 0 & c_{22} \end{bmatrix} - \lambda \begin{bmatrix} f_{11} & f_{12} \\ 0 & f_{22} \end{bmatrix}$$
with eigenvalues $\lambda_1 = c_{11}/f_{11}$ and $\lambda_2 = c_{22}/f_{22}$, and we swap the eigenvalues to get
$$\widehat{C} - \lambda \widehat{F} = \begin{bmatrix} \hat{c}_{11} & \hat{c}_{12} \\ \hat{c}_{21} & \hat{c}_{22} \end{bmatrix} - \lambda \begin{bmatrix} \hat{f}_{11} & \hat{f}_{12} \\ \hat{f}_{21} & \hat{f}_{22} \end{bmatrix}.$$

We can assume without loss of generality that $|\lambda_1| \geq |\lambda_2|$. The swapping operation is a unitary equivalence requiring four matrix multiplications, and such operations are backward stable [35], but there is one thing we have to check. The core Q is designed so that $Q^*(FZ)$ has a zero in the $(2,1)$ position, i.e., $\hat{f}_{21} = 0$. This automatically creates a zero in the $(2,1)$ position of $Q^*(CZ)$ (i.e., $\hat{c}_{21} = 0$) because CZe_1 and FZe_1 are proportional. This is true in exact arithmetic. We just need to check that in floating-point arithmetic the computed entry $\hat{c}_{21}$ is small enough that backward stability is not compromised by setting it to zero. For this it suffices that its magnitude be no bigger than a modest multiple of $u\|C\|$.

The swapping operation begins with the computation of the first row of $H = f_{22}C - c_{22}F$, which we will call w:
$$w = [\, f_{22}c_{11} - c_{22}f_{11} \quad f_{22}c_{12} - c_{22}f_{12} \,].$$

In floating-point arithmetic we get
$$\mathrm{fl}(w) = [\, f_{22}c_{11}(1+\varepsilon_1) - c_{22}f_{11}(1+\varepsilon_2) \quad f_{22}c_{12}(1+\varepsilon_3) - c_{22}f_{12}(1+\varepsilon_4) \,], \quad (2.3.1)$$

where each ε_i is the result of two roundoff errors, a multiplication and a subtraction mapped back to the product terms, and therefore satisfies $|\varepsilon_i| \leq 2u + O(u^2)$. Again we will use the abbreviation $|\varepsilon_i| \lesssim u$ to mean that $|\varepsilon_i|$ is no bigger than a modest constant times u.

The next step is to compute Z such that $wZ = [\, 0 \quad * \,]$. In practice we do this using $\mathrm{fl}(w)$ and make additional roundoff errors in the computation (2.1.16). We get $\widetilde{Z} = \mathrm{fl}(Z)$ satisfying
$$\widetilde{Z}e_1 = \tilde{x} = \tilde{\gamma}^{-1} \begin{bmatrix} -\mathrm{fl}(w_2)(1+\varepsilon_5) \\ \mathrm{fl}(w_1)(1+\varepsilon_6) \end{bmatrix}. \quad (2.3.2)$$

2.3. Backward stability in the case $m = k = 1$

Here $\tilde{\gamma} = \|\operatorname{fl}(w)\|$. In our discussion of the procedure (2.1.16) we noted that the quantities are computed to high relative accuracy. Therefore $|\varepsilon_5| \lesssim u$ and $|\varepsilon_6| \lesssim u$.

The vector $\tilde{x}$ defined by (2.3.2) is our computed (and normalized) version of a right eigenvector associated with eigenvalue λ_2. For later use we wish to show that $\tilde{x}$ is exactly an eigenvector of a slightly perturbed pencil. Thus we seek $\tilde{c}_{11}, \tilde{c}_{22}, \tilde{f}_{11}$, and $\tilde{f}_{22}$ such that

$$\left(\tilde{f}_{22}\begin{bmatrix} \tilde{c}_{11} & c_{12} \\ & \tilde{c}_{22} \end{bmatrix} - \tilde{c}_{22}\begin{bmatrix} \tilde{f}_{11} & f_{12} \\ & \tilde{f}_{22} \end{bmatrix}\right)\begin{bmatrix} \tilde{x}_1 \\ \tilde{x}_2 \end{bmatrix} = \begin{bmatrix} 0 \\ 0 \end{bmatrix}. \tag{2.3.3}$$

Notice that we are not going to back any of the error onto c_{12} or f_{12}. This equation is equivalent to

$$(\tilde{f}_{22}\tilde{c}_{11} - \tilde{c}_{22}\tilde{f}_{11})\tilde{x}_1 + (\tilde{f}_{22}c_{12} - \tilde{c}_{22}f_{12})\tilde{x}_2 = 0.$$

Filling in the values of $\tilde{x}_1$ and $\tilde{x}_2$ from (2.3.2) and (2.3.1), we can check that this equation holds if we make the assignments

$$\tilde{c}_{11} = c_{11}\frac{(1+\varepsilon_1)(1+\varepsilon_6)}{(1+\varepsilon_3)(1+\varepsilon_5)}, \qquad \tilde{c}_{22} = c_{22}(1+\varepsilon_4)(1+\varepsilon_5),$$

$$\tilde{f}_{11} = f_{11}\frac{(1+\varepsilon_2)(1+\varepsilon_6)}{(1+\varepsilon_4)(1+\varepsilon_5)}, \qquad \tilde{f}_{22} = f_{22}(1+\varepsilon_3)(1+\varepsilon_5).$$

Clearly $|\tilde{c}_{ii} - c_{ii}| \lesssim u|c_{ii}|$ and $|\tilde{f}_{ii} - f_{ii}| \lesssim u|f_{ii}|$ for $i = 1, 2$. Equation (2.3.3) can be written more compactly as

$$\tilde{f}_{22}\tilde{C}\tilde{x} = \tilde{c}_{22}\tilde{F}\tilde{x}. \tag{2.3.4}$$

Thus $\tilde{x}$ is an eigenvector of the perturbed pencil $\tilde{C} - \lambda \tilde{F}$ associated with eigenvalue $\tilde{\lambda}_2 = \tilde{c}_{22}/\tilde{f}_{22}$. We also write

$$\tilde{C} = C + \delta C \quad \text{and} \quad \tilde{F} = F + \delta F_1, \tag{2.3.5}$$

with δC and δF_1 diagonal matrices satisfying

$$\|\delta C\| \lesssim u\|C\| \quad \text{and} \quad \|\delta F_1\| \lesssim u\|F\|.$$

Finally we compute Q. In exact arithmetic Q is constructed so that its first column is proportional to FZe_1. In practice, instead of FZe_1 we use

$$\breve{y} = \operatorname{fl}\left(F\tilde{Z}e_1\right) = \operatorname{fl}(F\tilde{x}) = \tilde{\gamma}^{-1}\begin{bmatrix} f_{11}\tilde{x}_1(1+\varepsilon'_1) + f_{12}\tilde{x}_2(1+\varepsilon'_2) \\ f_{22}\tilde{x}_2(1+\varepsilon'_3) \end{bmatrix},$$

where $|\varepsilon'_i| \lesssim u$ for $i = 1, 2, 3$. The computed version of Q is $\tilde{Q} = \operatorname{fl}(Q)$ satisfying

$$\tilde{Q}e_1 = \breve{\zeta}^{-1}\begin{bmatrix} \breve{y}_1(1+\varepsilon'_4) \\ \breve{y}_2(1+\varepsilon'_5) \end{bmatrix},$$

where $\breve{\zeta} = \|\breve{y}\|$, and ε'_4 and ε'_5 are due to the tiny roundoff errors in the calculation (2.1.16).

For our analysis we need to establish that there is a slightly perturbed matrix

$$\hat{F} = F + \delta F_2 = \begin{bmatrix} \hat{f}_{11} & f_{12} \\ & \hat{f}_{22} \end{bmatrix}$$

such that $\tilde{Q}^*\hat{F}\tilde{Z}$ has an exact zero in the $(2,1)$ position. This just means that $\tilde{y} = \tilde{Q}e_1$ is exactly proportional to $\hat{F}\tilde{Z}e_1 = \hat{F}\tilde{x}$. It is easy to check that the choice

$$\hat{f}_{11} = f_{11}\frac{(1+\varepsilon'_1)}{(1+\varepsilon'_2)}, \qquad \hat{f}_{22} = f_{22}\frac{(1+\varepsilon'_3)(1+\varepsilon'_5)}{(1+\varepsilon'_2)(1+\varepsilon'_4)}$$

does the trick. Clearly $|\hat{f}_{11} - f_{11}| \lesssim u|f_{11}|$ and $|\hat{f}_{22} - f_{22}| \lesssim u|f_{22}|$, and δF_2 is a diagonal matrix satisfying $\|\delta F_2\| \lesssim u\|F\|$.

Our final computed results are $\mathrm{fl}(\tilde{Q}^* C \tilde{Z})$ and $\mathrm{fl}(\tilde{Q}^* F \tilde{Z})$. We have to show that the $(2,1)$ entries of these matrices are small enough that we can set them to zero without compromising backward stability. The "F" part is routine. Focusing on the $(2,1)$ entry, we have

$$e_2^T \mathrm{fl}(\tilde{Q}^* F \tilde{Z}) e_1 = e_2^T \tilde{Q}^* F \tilde{Z} e_1 + e_2^T E_1 e_1,$$

where E_1 is the matrix of roundoff errors incurred in multiplying the three matrices together and satisfies $\|E_1\| \lesssim u \|\tilde{Q}\| \|F\| \|\tilde{Z}\|$, i.e., $\|E_1\| \lesssim u \|F\|$. The remaining term is

$$e_2^T \tilde{Q}^* F \tilde{Z} e_1 = e_2^T \tilde{Q}^* \hat{F} \tilde{Z} e_1 - e_2^T \tilde{Q}^* \delta F_2 \tilde{Z} e_1.$$

The first term on the right-hand side is exactly zero by construction. The second is bounded above by $\|\delta F_2\| \lesssim u \|F\|$. This takes care of the "F" part.

The "C" part (the important part) is more delicate. We have

$$e_2^T \mathrm{fl}(\tilde{Q}^* C \tilde{Z}) e_1 = e_2^T \tilde{Q}^* C \tilde{Z} e_1 + e_2^T E_2 e_1,$$

where E_2 is the matrix of roundoff errors incurred in multiplying the three matrices together and satisfies $\|E_2\| \lesssim u \|C\|$. The remaining term is

$$e_2^T \tilde{Q}^* C \tilde{Z} e_1 = e_2^T \tilde{Q}^* \tilde{C} \tilde{Z} e_1 - e_2^T \tilde{Q}^* \delta C \tilde{Z} e_1.$$

The second term on the right-hand side is bounded above by $\|\delta C\| \lesssim u \|C\|$, so now we can just focus on the other term. Here we make use of (2.3.4), which can be written as $\tilde{C} \tilde{Z} e_1 = (\tilde{c}_{22}/\tilde{f}_{22}) \tilde{F} \tilde{Z} e_1$. Thus

$$e_2^T \tilde{Q}^* \tilde{C} \tilde{Z} e_1 = \frac{\tilde{c}_{22}}{\tilde{f}_{22}} e_2^T \tilde{Q}^* \tilde{F} \tilde{Z} e_1 = \frac{\tilde{c}_{22}}{\tilde{f}_{22}} e_2^T \tilde{Q}^* \hat{F} \tilde{Z} e_1 + \frac{\tilde{c}_{22}}{\tilde{f}_{22}} e_2^T \tilde{Q}^* (\delta F_1 - \delta F_2) \tilde{Z} e_1.$$

The term containing $\hat{F}$ is zero by construction, so now we just need to concentrate on the other term. Let $\delta F = \delta F_1 - \delta F_2$. From the definitions of δF_1 and δF_2 we see that

$$\delta F = \begin{bmatrix} \varepsilon_1'' f_{11} & 0 \\ 0 & \varepsilon_2'' f_{22} \end{bmatrix},$$

where $|\varepsilon_i''| \lesssim u$ for $i = 1, 2$. Moreover $\dfrac{\tilde{c}_{22}}{\tilde{f}_{22}} = \dfrac{c_{22}}{f_{22}}(1 + \varepsilon_3'')$ for some tiny ε_3''. We also use our assumption $|\lambda_1| \geq |\lambda_2|$ to deduce that $|f_{11} c_{22}/f_{22}| \leq |c_{11}|$. Thus

$$|(\tilde{c}_{22}/\tilde{f}_{22}) \delta F| = |1 + \varepsilon_3''| \begin{bmatrix} |\varepsilon_1'' f_{11} c_{22}/f_{22}| & \\ & |\varepsilon_2'' c_{22}| \end{bmatrix} \leq |1 + \varepsilon_3''| \begin{bmatrix} |\varepsilon_1'' c_{11}| & \\ & |\varepsilon_2'' c_{22}| \end{bmatrix},$$

so

$$\|(\tilde{c}_{22}/\tilde{f}_{22}) \delta F\| \lesssim u \|C\|.$$

We conclude that our one remaining term, which is $(\tilde{c}_{22}/\tilde{f}_{22}) e_2^T \tilde{Q}^* (\delta F) \tilde{Z} e_1$, satisfies

$$|(\tilde{c}_{22}/\tilde{f}_{22}) e_2^T \tilde{Q}^* (\delta F) \tilde{Z} e_1| \lesssim u \|C\|.$$

This completes the "C" part.

We have demonstrated that

$$|e_2^T \mathrm{fl}(\tilde{Q}^* C \tilde{Z}) e_1| \lesssim u \|C\| \quad \text{and} \quad |e_2^T \mathrm{fl}(\tilde{Q}^* F \tilde{Z}) e_1| \lesssim u \|F\|,$$

so we can set these numbers to zero without compromising backward stability. The $\lesssim$ symbols hide constants, but these constants are not too large due to the small total number of operations required by the swap.

Chapter 3
Krylov Processes

3.1 ▪ Krylov subspaces

In Chapter 1 we suggested a general approach to solving eigenvalue problems: reduce the matrix to a condensed form (e.g., Hessenberg), then iterate to get the eigenvalues. At each step we do a similarity transformation. This approach is feasible for small- to medium-sized problems with n not exceeding a few thousand. For larger problems, other methods are required. The Krylov subspace methods that are discussed in this chapter are among the most important.[6]

Important as they are, Krylov processes are not the main focus of this book. Their presence here is meant as motivation for the following chapter on pole-swapping algorithms. Accordingly this presentation is just a brief sketch. We will have more to say about Krylov processes in Chapter 6, in which we discuss filtering by implicit restarts.

In most applications involving very large matrices, the matrices are also *sparse*, meaning that only a small fraction of their entries are nonzero. Such matrices can be stored compactly using a data structure that stores only the nonzero entries; one simple scheme would be a list of ordered triples (a_{ij}, i, j) that records the numerical value a_{ij} and coordinates (i, j) of each nonzero element. This can save a lot of space. For example, suppose the dimension of the matrix is $n = 10^6$ but only about 10 elements in each row are nonzero. If we store the matrix in the conventional way, we have to store $n^2 = 10^{12}$ numbers. If we store only the nonzeros, we only have to store about $20n = 2 \times 10^7$ numbers, a huge saving in space. In this calculation we have thrown in an extra factor 2 to account for the fact that along with each real or complex nonzero entry a_{ij}, we also have to store the two integers i and j.

If we want to exploit sparseness, we cannot do similarity transforms, as that would generally destroy the sparseness. We will assume that there is one thing that we can do, and that is to multiply A by an arbitrary vector x to obtain a new vector Ax. Suppose the sparse matrix A is stored as a list, as described above. Then, for any vector x, it is possible to compute Ax with one pass through the list (Exercise 1). The flop count for this operation is twice the number of nonzeros in A, so it's cheap if A is sparse.

If we can compute Ax, then we can multiply A times Ax to obtain $A^2 x$. We can then multiply A by $A^2 x$ to get $A^3 x$, and so on. For $k = 1, 2, 3, \ldots$ we define the kth *Krylov subspace* $\mathcal{K}_k(A, x)$ by

$$\mathcal{K}_k(A, x) = \operatorname{span}\{x, Ax, A^2 x, \ldots, A^{k-1} x\}.$$

---

[6]Krylov subspaces have other uses as well. For example, they play a role in the study of Francis's and other bulge-chasing algorithms in [69], and are crucial to the convergence theory of Francis's algorithm laid out in [70, 71]. In the present book, in our discussions of the convergence of pole-swapping algorithms, we introduce *rational* Krylov subspaces. See Section 4.7.

Krylov subspace methods search for good approximations to eigenvalues/eigenvectors of A within Krylov subspaces.

The Krylov subspaces associated with a given A and x are clearly nested:

$$\mathcal{K}_1(A, x) \subseteq \mathcal{K}_2(A, x) \subseteq \mathcal{K}_3(A, x) \subseteq \cdots.$$

Generically the space $\mathcal{K}_m(A, x)$ will have dimension m with a basis x, Ax, $A^2 x$, ..., $A^{m-1} x$, at least for small enough m. This holds up until we run into an invariant subspace. The reader can easily check that, for $j = 1, \ldots, m$, $\mathcal{K}_j(A, x)$ has dimension j as long as none of the spaces $\mathcal{K}_j(A, x)$ is invariant under A, except possibly $\mathcal{K}_m(A, x)$. If $\mathcal{K}_m(A, x)$ is invariant, then $\mathcal{K}_m(A, x) = \mathcal{K}_{m+1}(A, x) = \mathcal{K}_{m+2}(A, x) = \cdots$.

In our discussion of Krylov subspaces, we will generally make the unstated assumption that the spaces we are dealing with are not invariant. If we are lucky enough to get to an invariant subspace $\mathcal{K}_m(A, x)$ for some $m \ll n$, that's good news, as it allows us to extract m eigenvalues and corresponding vectors of A. This will become clear in our discussion below.

For future reference we note that Krylov subspaces are shift invariant: for any shift ρ we have, for any m,

$$\mathcal{K}_m(A - \rho I, x) = \mathcal{K}_m(A, x).$$

More generally we have the following result.

Proposition 3.1.1. *Let* $\rho_1, \ldots, \rho_{m-1}$ *be any* $m - 1$ *complex shifts. Then, for* $j = 1, \ldots, m - 1$, $\mathcal{K}_j(A, x) =$

$$\operatorname{span}\{x, (A - \rho_1 I)x, (A - \rho_2 I)(A - \rho_1 I)x, \ldots, (A - \rho_{j-1} I) \cdots (A - \rho_1 I)x\}.$$

The proof is left as an exercise for the reader.

Exercises 3.1

1. (a) Suppose the sparse matrix A is stored as a list of the nonzero entries (a_{ij}, i, j), as described in this section. Show that for any vector x, the product Ax can be computed with one pass through the list. (Start with 0 in the vector that is going to become Ax, and add in one term for each list entry.)

    (b) Deduce that the flop count for this operation is two times the number of nonzeros in A.

2. Consider a sequence of Krylov subspaces $\mathcal{K}_1(A, x)$, $\mathcal{K}_2(A, x)$, ...

    (a) Show that $\mathcal{K}_j(A, x)$ is invariant under A if and only if $A^j x$ is a linear combination of x, Ax, ..., $A^{j-1} x$.

    (b) Show that, for $j = 1, \ldots, m$, $\mathcal{K}_j(A, x)$ has dimension j as long as none of the spaces $\mathcal{K}_j(A, x)$ are invariant under A for $j = 1, \ldots, m - 1$.

    (c) Show that if $\mathcal{K}_j(A, x)$ is invariant, then $\mathcal{K}_j(A, x) = \mathcal{K}_{j+1}(A, x) = \mathcal{K}_{j+2}(A, x) = \cdots$.

3. Prove Proposition 3.1.1.

3.2 ▪ The Arnoldi process

The simplest and most important Krylov algorithm is the *Arnoldi process* [2].

In order to work with a Krylov subspace or any subspace, we need a basis. The space $\mathcal{K}_j(A, x)$ has the obvious basis x, Ax, A^2x, ..., $A^{j-1}x$, but this basis becomes increasingly ill conditioned as j increases. We can obtain a much better basis by orthonormalizing x, Ax, ..., $A^{j-1}x$ by the Gram–Schmidt process. In principle this will yield an orthonormal basis q_1, ..., q_j such that for $i = 1, \ldots, j$, $\text{span}\{q_1, \ldots, q_i\} = \mathcal{K}_i(A, x)$. If we want to continue the process, we compute $A^j x$ and then orthogonalize it against $q_1, \ldots, q_j$.

The Arnoldi process is a variant of Gram–Schmidt that is more stable and efficient. It starts out, like Gram–Schmidt, by computing $q_1 = x/\|x\|$. After $j-1$ steps we have orthonormal q_1, ..., q_j such that, for $i = 1, \ldots, j$, $\text{span}\{q_1, \ldots, q_i\} = \mathcal{K}_i(A, x)$. We continue the process with Aq_j instead of $A^j x$. This is the only difference. Thus step $j+1$ computes Aq_j then

$$\hat{q}_{j+1} = Aq_j - \sum_{i=1}^{j} q_i h_{ij}, \tag{3.2.1}$$

where each h_{ij} is chosen so that $\hat{q}_{j+1}$ is orthogonal to q_i:

$$h_{ij} = \langle Aq_j, q_i \rangle, \qquad i = 1, \ldots, j. \tag{3.2.2}$$

The reader can easily check that $\hat{q}_{j+1} = 0$ if and only if $\text{span}\{q_1, \ldots, q_j\}$ is invariant under A. This is a good result, as it allows us to extract j eigenpairs from the invariant subspace. If $\hat{q}_{j+1} \neq 0$, we finish the step by normalizing it:

$$h_{j+1,j} = \|\hat{q}_{j+1}\| > 0, \qquad q_{j+1} = \hat{q}_{j+1}/h_{j+1,j}. \tag{3.2.3}$$

Notice that with this variant of Gram–Schmidt, we never have to compute or store $A^2 x$, $A^3 x$, ... Instead it suffices to compute $q_1, q_2, \ldots$. We also save the coefficients h_{ij}. In a moment we will see what we can do with this information.[7] First notice that by rearranging (3.2.1) and using (3.2.3) we get the representation

$$Aq_j = \sum_{i=1}^{j+1} q_i h_{ij}, \qquad j = 1, 2, 3, \ldots. \tag{3.2.4}$$

This holds for $j = 1, \ldots, m$ if x, Ax, ..., $A^m x$ are linearly independent. We generate q_1, ..., q_{m+1} and accompanying coefficients h_{ij}. Define

$$Q_m = \begin{bmatrix} q_1 & q_2 & \cdots & q_m \end{bmatrix} \in \mathbb{C}^{n \times m},$$

which also defines $Q_{m+1} \in \mathbb{C}^{n \times (m+1)}$. Let

$$H_{m+1,m} = \begin{bmatrix} h_{11} & h_{12} & \cdots & h_{1,m-1} & h_{1,m} \\ h_{21} & h_{22} & \cdots & h_{2,m-1} & h_{2,m} \\ 0 & h_{32} & \cdots & h_{3,m-1} & h_{3,m} \\ \vdots & & \ddots & & \vdots \\ 0 & & & h_{m,m-1} & h_{m,m} \\ 0 & 0 & \cdots & 0 & h_{m+1,m} \end{bmatrix} \in \mathbb{C}^{(m+1),m}. \tag{3.2.5}$$

---

[7] In this brief sketch we have omitted the fact that to guarantee stability it is often advisable to do the orthogonalization twice. That is, once $\hat{q}_{j+1}$ has been computed by (3.2.1), a second orthogonalization step should be applied to $\hat{q}_{j+1}$ to obtain a refined $\hat{q}_{j+1}$.

This is a nonsquare upper Hessenberg matrix with positive entries on the subdiagonal. From (3.2.4) we see immediately that

$$AQ_m = Q_{m+1}H_{m+1,m} \tag{3.2.6}$$

because the jth column of this equation is exactly (3.2.4).

We can derive a useful variant of (3.2.6) as follows. Let H_m be the $m \times m$ matrix obtained from $H_{m+1,m}$ by deleting the last row. Then

$$Q_{m+1}H_{m+1,m} = Q_m H_m + q_{m+1} \begin{bmatrix} 0 & \cdots & 0 & h_{m+1,m} \end{bmatrix} = Q_m H_m + q_{m+1} h_{m+1,m} e_m^T.$$

Thus, from (3.2.6) we obtain

$$AQ_m = Q_m H_m + q_{m+1} h_{m+1,m} e_m^T. \tag{3.2.7}$$

We think of the last term as a remainder. Equation (3.2.7) holds if we are able to complete m steps of the Arnoldi process. If, on the other hand, we are able to complete $m-1$ steps but get $\hat{q}_{m+1} = 0$ on the mth step, then we have

$$AQ_m = Q_m H_m \tag{3.2.8}$$

with no remainder term. In this case $\mathcal{R}(Q_m)$, which is the same thing as $\text{span}\{q_1, \ldots, q_m\}$ and $\mathcal{K}_m(A, x)$, is invariant under A by Proposition 1.1.3. The reader can check that if v is an eigenvector of H_m with eigenvalue λ, then $Q_m v$ is an eigenvector of A with eigenvalue λ. Thus every eigenvalue of H_m is also an eigenvalue of A, and we can get eigenvectors too. This begins to show how the Arnoldi process can be used to compute eigenvalues/vectors of A.

Of course, we are not usually so lucky as to hit an invariant subspace exactly. Nevertheless we can still look in H_m for approximations to eigenvalues of A. Suppose v is an eigenvector of H_m (with $\|v\| \approx 1$) associated with eigenvalue λ. Multiplying (3.2.7) by v we obtain

$$A(Q_m v) = \lambda (Q_m v) + q_{m+1} h_{m+1,m} v_m,$$

where v_m denotes the mth component of v. The norm of the remainder term is clearly $|h_{m+1,m} v_m|$. If this number is tiny, then we have a good approximate eigenpair in the sense that the residual $\|A(Q_m v) - \lambda(Q_m v)\|$ is tiny. If $|h_{m+1,m}|$ is tiny, then all the eigenvalues of H_m are good approximations, but even if it isn't there may be some that are good approximations, namely those for which $|v_m|$ is tiny. Normally the eigenvalues of A that are approximated best are on the periphery of the spectrum of A [70]. (But see also [43, 44, 12].)

Typically we run the Arnoldi process for m steps, where $m \ll n$. Thus the storage space required for $q_1, \ldots, q_m$ is not excessive, nor is that for the $m \times m$ matrix H_m. This matrix is already in upper Hessenberg form, so we can easily compute its eigenvalues/vectors by Francis's algorithm.

Although we would not normally do it in practice, it is interesting to consider what would happen if we ran the Arnoldi process for n steps. It would conclude with $\hat{q}_{n+1} = 0$ because the vectors $q_1, \ldots, q_n$ fill all of space, that is, $\text{span}\{q_1, \ldots, q_n\} = \mathbb{C}^n$. Then we would have

$$AQ_n = Q_n H_n,$$

the case $m = n$ in (3.2.8). Both Q_n and H_n are $n \times n$, Q_n is unitary, and H_n is upper Hessenberg. Since $A = Q_n H_n Q_n^{-1}$, we see that *the Arnoldi process, when carried to completion, performs a similarity transformation of A to upper Hessenberg form*. This means that if we only carry out $m < n$ steps, arriving at $AQ_m = Q_{m+1} H_{m+1,m}$ (3.2.6), we are doing a partial similarity transformation or "the beginning of" a similarity transformation.

The shift-and-invert strategy

We have noted that the Arnoldi process is good at approximating the eigenvalues on the periphery of the spectrum of A. What if we want to compute eigenvalues that are not out on the edge? One possibility is to pick a target shift τ and apply the Arnoldi process to $(A - \tau I)^{-1}$ instead of A. This is the *shift-and-invert strategy*. For each eigenvalue λ of A, the quantity $\mu = (\lambda - \tau)^{-1}$ is an eigenvalue of $(A - \tau I)^{-1}$. For each eigenvalue μ of $(A - \tau I)^{-1}$ that we compute, $\lambda = \tau + 1/\mu$ is an eigenvalue of A. This allows us to find those eigenvalues of A that are nearest to τ, as these correspond to the peripheral eigenvalues of $(A - \tau I)^{-1}$.[8]

This procedure requires greater computational effort than standard Arnoldi; at step j we have to compute the vector $z = (A - \tau I)^{-1} q_j$ instead of Aq_j. Of course we do not actually compute the (nonsparse) inverse matrix. Instead we solve the linear system $(A - \tau I)z = q_j$ using a sparse LU decomposition [70]. Notice that we have to compute this decomposition just once at the beginning; then we can reuse it at each step. If we can afford to do this in terms of computational time and storage space, we can shift and invert.

Exercises 3.2

1. Suppose the orthonormal vectors $q_1, \ldots, q_m$ are generated by the Arnoldi process with starting vector x. Prove by induction that $\operatorname{span}\{q_1, \ldots, q_j\} = \mathcal{K}_j(A, x)$ for $j = 1, \ldots, m$.

2. Consider the computation (3.2.1) using the coefficients (3.2.2).

   (a) Show that $\hat{q}_{j+1}$ is orthogonal to $q_1, \ldots, q_j$.

   (b) Show that $\hat{q}_{j+1} = 0$ if and only if $\mathcal{K}_j(A, x)$ is invariant under A.

3.3 ▪ Generalized Arnoldi process

To move our story along, we introduce a modest generalization. The standard Arnoldi process continues with the vector Aq_j at the jth step. Instead of q_j we could use a different continuation vector, a linear combination of $q_1, \ldots, q_j$:

$$\tilde{q}_j = \sum_{i=1}^{j} q_i k_{ij}, \quad k_{jj} \neq 0. \tag{3.3.1}$$

Then we continue with

$$\hat{q}_{j+1} = A\tilde{q}_j - \sum_{i=1}^{j} q_i h_{ij}, \tag{3.3.2}$$

where the coefficients h_{ij} are chosen so that $\hat{q}_{j+1}$ is orthogonal to $q_1, \ldots, q_j$:

$$h_{ij} = \langle A\tilde{q}_j, q_i \rangle, \quad i = 1, \ldots, j. \tag{3.3.3}$$

Then we normalize:

$$h_{j+1,j} = \|\hat{q}_{j+1}\|, \quad q_{j+1} = \hat{q}_{j+1}/h_{j+1,j}. \tag{3.3.4}$$

If we now substitute (3.3.1) and (3.3.4) into (3.3.2) and do some simple manipulations, we obtain

$$A\left(\sum_{i=1}^{j} q_i k_{ij}\right) = \sum_{i=1}^{j+1} q_i h_{ij}. \tag{3.3.5}$$

---

[8] This technique can also be applied to the generalized eigenvalue problem $Av = \lambda Bv$. In this case we use $(A - \tau B)^{-1}$ in place of $(A - \tau I)^{-1}$.

Now we can get some matrix representations of the procedure. Suppose we have done m steps of this generalized Arnoldi process to obtain $q_1, \ldots, q_{m+1}$ and coefficients h_{ij} and k_{ij}. Define $H_{m+1,m}$ by (3.2.5) as before, and let K_m denote the nonsingular, upper-triangular matrix built from the coefficients k_{ij} in the obvious way. Then we obtain

$$AQ_m K_m = Q_{m+1} H_{m+1,m}. \tag{3.3.6}$$

This is so because, for $j = 1, \ldots, m$, the jth column of (3.3.6) is exactly (3.3.5). This can also be written with a square H_m and a remainder term:

$$AQ_m K_m = Q_m H_m + q_{m+1} h_{m+1,m} e_m^T. \tag{3.3.7}$$

From these equations we can extract approximate eigenvalues in a way similar to but slightly different from the standard Arnoldi process. We consider first the invariant case. If $\mathrm{span}\{q_1, \ldots, q_m\}$ is invariant under A, then (3.3.6) and (3.3.7) both simplify to

$$AQ_m K_m = Q_m H_m.$$

From this we can easily check that if λ is an eigenvalue of the pair (H_m, K_m) with eigenvector v, i.e., $H_m v = \lambda K_m v$, then $Q_m K_m v$ is an eigenvector of A with eigenvalue λ. Thus every eigenvalue of (H_m, K_m) is an eigenvalue of A.

We are usually not so lucky as to hit an invariant subspace exactly. Nevertheless, we can still compute the eigenvalues of (H_m, K_m) and expect some of them to be good approximations of eigenvalues of A. Since H_m is upper Hessenberg and K_m is upper triangular, we have an algorithm that can compute the eigenvalues, namely the Moler–Stewart QZ algorithm.

Exercises 3.3

1. Explain why the condition $k_{jj} \neq 0$ is needed in (3.3.1).

3.4 • Rational Krylov process

The *rational Krylov process* or *rational Arnoldi process* was introduced by Axel Ruhe [52] and developed in [53, 54, 55] (see also [34]). It resembles the shift-and-invert strategy mentioned in Section 3.2, except that we allow a different shift at every step. In this context the shifts are generally called *poles*. If we use σ_j as the pole at step j, we have

$$q_{j+1} k_{j+1,j} = (A - \sigma_j I)^{-1} q_j - \sum_{i=1}^{j} q_i k_{ij},$$

where the k_{ij} are chosen so that $q_1, \ldots, q_{j+1}$ are orthonormal. We have shifted the notation from h_{ij}, as in (3.2.1), to k_{ij}. This turns out to be a convenient change because we are now dealing with an inverse matrix.

More generally we can use a continuation vector $\tilde{q}_j = \sum_{i=1}^{j} q_i c_{ij}$, $(c_{jj} \neq 0)$, as in (3.3.1). Then we have

$$q_{j+1} k_{j+1,j} = (A - \sigma_j I)^{-1} \tilde{q}_j - \sum_{i=1}^{j} q_i k_{ij}. \tag{3.4.1}$$

If we multiply through by $A - \sigma_j I$ and do some simple manipulations, we find that

$$A \left(\sum_{i=1}^{j+1} q_i k_{ij} \right) = \sum_{i=1}^{j} q_i (c_{ij} + k_{ij} \sigma_j) + q_{j+1} k_{j+1,j} \sigma_j. \tag{3.4.2}$$

3.4. Rational Krylov process

Say we do m steps of this process. Define an $(m+1) \times m$ upper Hessenberg matrix $H_{m+1,m}$ by

$$h_{ij} = \begin{cases} c_{ij} + k_{ij}\sigma_j & \text{if } i \leq j, \\ k_{ij}\sigma_j & \text{if } i = j+1, \\ 0 & \text{if } i > j+1. \end{cases} \quad (3.4.3)$$

Then we can rewrite (3.4.2) in the simpler form

$$A\left(\sum_{i=1}^{j+1} q_i k_{ij}\right) = \sum_{i=1}^{j+1} q_i h_{ij}, \quad (3.4.4)$$

and from this we immediately obtain

$$AQ_{m+1}K_{m+1,m} = Q_{m+1}H_{m+1,m}. \quad (3.4.5)$$

Again this is true because the jth column of (3.4.5) is exactly (3.4.4). Both $H_{m+1,m}$ and $K_{m+1,m}$ are upper Hessenberg. The poles are encoded in this pair in the form

$$\sigma_j = h_{j+1,j}/k_{j+1,j}, \quad j = 1,\ldots, m, \quad (3.4.6)$$

as is clear from (3.4.3).

As before, we can write (3.4.5) using square matrices K_m, H_m, and a remainder term:

$$AQ_m K_m = Q_m H_m + (q_{m+1}\sigma_m - Aq_{m+1})k_{m+1,m}e_m^T. \quad (3.4.7)$$

If $\mathcal{R}(Q_m)$ is invariant under A, all eigenvalues of the pair (H_m, K_m) will be eigenvalues of A. Even if it's not invariant, some of the eigenvalues of (H_m, K_m) may be good approximations to eigenvalues of A. Both H_m and K_m are upper Hessenberg, so we call this a *Hessenberg pair*. How should we compute the eigenvalues? Since neither of the matrices is triangular, we cannot directly apply the Moler–Stewart algorithm. In the next chapter we will introduce pole-swapping algorithms that operate on Hessenberg pencils and can therefore solve this problem directly.

Berljafa and Güttel [13] introduced methods for manipulating rational Krylov decompositions like (3.4.5). They showed how to alter the poles in the pair $(H_{m+1,m}, K_{m+1,m})$ and move them around. The moves introduced there are exactly the moves we will use in our pole-swapping algorithms in Chapter 4.

Exercises 3.4

1. Show the details of the computation (3.4.1). Do we actually compute the matrix $(A - \sigma_j I)^{-1}$? What is the formula for the coefficients k_{ij}, including $k_{j+1,j}$, that guarantee that $q_1, \ldots, q_{j+1}$ are orthonormal? How do we detect an invariant subspace?

2. Express (3.4.3) as a matrix equation: $H_{m+1,m} = \cdots$.

3. (Insertion of an infinite pole) At any point in the rational Krylov process we can insert an ordinary Arnoldi step, multiplying by A instead of $(A - \sigma I)^{-1}$. Suppose we do this at step j, that is, we use (3.3.2) instead of (3.4.1). How does this change equation (3.4.5)? Show that we get $h_{j+1,j} \neq 0$ and $k_{j+1,j} = 0$. This means we have inserted an infinite pole, according to formula (3.4.6).

4. In the previous exercise we introduced the idea that we can apply both positive and negative powers in the rational Krylov process. This suggests that we might benefit from

applying both positive and negative powers in the same step. For example, we might try $(A - \rho I)(A - \sigma I)^{-1}$, where $\rho \neq \sigma$. Show that

$$(A - \rho I)(A - \sigma I)^{-1} = I + (\sigma - \rho)(A - \sigma I)^{-1}.$$

This shows that the factor $A - \rho I$ contributes nothing.

5. In some developments of the rational Krylov method, for example, [34], the operator $(I - \sigma_j^{-1} A)^{-1} A$ is used instead of $(A - \sigma_j I)^{-1}$ in (3.4.1).

   (a) Show (by induction) that both operators result in the same space

   $$\text{span}\{q_1, \ldots, q_j, q_{j+1}\}.$$

   Hints: (i) This is true regardless of how $\tilde{q}_j$ is chosen, but to avoid irrelevant complications, just prove the result in the special case $\tilde{q}_j = q_j$. (ii) This exercise is related to the previous one.

   (b) If the operator $(I - \sigma_j^{-1} A)^{-1} A$ is used, what happens in the case $\sigma_j = \infty$?

6. Show that $A - \rho I$ and $(A - \sigma I)^{-1}$ commute.

Chapter 4
Pole-Swapping Algorithms

This chapter draws heavily from [21] and [25].

The rational Krylov algorithm produces Hessenberg pairs. In this chapter we will introduce pole-swapping algorithms that compute the eigenvalues/vectors of such a pair. As we mentioned in Chapter 2, pole swapping turns out to be a matter of interchanging blocks of eigenvalues. In this chapter we look at the simplest case, in which the blocks are 1×1. Larger blocks will be discussed in Chapter 7.

4.1 ▪ Operations on Hessenberg pairs

The pair (A, B) is called a *Hessenberg pair* if both A and B are Hessenberg matrices. If $a_{j+1,j} = 0 = b_{j+1,j}$ for some j, we can immediately split the eigenvalue problem into two smaller problems. We therefore eliminate that case from further consideration. The ratios $a_{j+1,j}/b_{j+1,j}$, $j = 1, \ldots, n-1$, are called the *poles* of the Hessenberg pair. In the case $b_{j+1,j} = 0$, we have an infinite pole. The Hessenberg-triangular form is a special type of Hessenberg pair for which all of the poles are infinite.

Closely related to (A, B) is the *pole pair* (A_π, B_π) (or *pole pencil* $A_\pi - \lambda B_\pi$) obtained from (A, B) by deleting the first row and last column. The pole pencil is upper triangular, and its eigenvalues are obviously the poles of (A, B).

Following [25] and [13] we introduce two types of operations, or *moves*, both of which manipulate the poles in the pair. Let $\sigma_1 = a_{21}/b_{21}, \ldots, \sigma_{n-1} = a_{n,n-1}/b_{n,n-1}$ denote the poles of the Hessenberg pair (A, B).

Changing a pole at the top or bottom. (Type I move)

We can change the pole σ_1 to any value we want by applying a core transformation Q_1^* to the pencil on the left. Suppose we want to change σ_1 to ρ, say. Noting that only the first two entries of $(A - \rho B)e_1$ can be nonzero, we deduce that there is a Q_1 such that the second entry of $Q_1^*(A - \rho B)e_1$ is zero. In other words,

$$Q_1^*(A - \rho B)e_1 = \gamma e_1 \tag{4.1.1}$$

for some γ. If we then define $\hat{A} = Q_1^* A$ and $\hat{B} = Q_1^* B$, then $(\hat{A} - \rho \hat{B})e_1 = \gamma e_1$, which implies that $\hat{a}_{21} - \rho \hat{b}_{21} = 0$. This means that $\rho = \hat{a}_{21}/\hat{b}_{21}$ is the new first pole of $(\hat{A}, \hat{B})$. The other poles remain fixed, as they are untouched by the transformation. (The core Q_1 defined here is exactly the transformation that is used to initiate an iteration of the single-shift QZ algorithm with shift ρ.)

This operation fails only if $\hat{a}_{21} = 0 = \hat{b}_{21}$, yielding $\rho = 0/0$. This happens exactly when the first columns of A and B are proportional. But this is not such a failure after all, as it exposes $\hat{a}_{11}/\hat{b}_{11}$ as an eigenvalue of the pencil and allows us to deflate to a smaller problem by deleting the first row and column.

In summary, if we want to replace pole σ_1 by ρ, we will either succeed in doing so or get a deflation of an eigenvalue.

Remark 4.1.1. *When we write something like $A - \rho B$ here and elsewhere, this should be viewed as shorthand for $\beta A - \alpha B$ where α and β are any scalars for which $\rho = \alpha/\beta$. As a practical matter this allows us to use modest sized α and β even when ρ is very large, and in particular it allows us to implement the case $\rho = \infty$ by taking $\beta = 0$. Thus $A - \rho B$ really stands for an equivalence class of operators, all of which are nonzero multiples of one another.*

The pole σ_{n-1} at the bottom can also be replaced by any other pole, say τ, by a similar procedure. We want to transform the pencil $A - \lambda B$ to $\hat{A} - \lambda \hat{B} = (A - \lambda B)Z_{n-1}$ with $\hat{a}_{n,n-1}/\hat{b}_{n,n-1} = \tau$. Noting that the row vector $e_n^T(A - \tau B)$ has nonzero entries only in its last two positions, we see that there must be a core transformation Z_{n-1} that maps it to a multiple of e_n^T, i.e., $e_n^T(A - \tau B)Z_{n-1} = \gamma e_n^T$ for some γ. This is the desired transformation, since it implies $e_n^T(\hat{A} - \tau \hat{B}) = \gamma e_n^T$, which is equivalent to $\hat{a}_{n,n-1}/\hat{b}_{n,n-1} = \tau$.

This fails only if $\hat{a}_{n,n-1} = 0 = \hat{b}_{n,n-1}$, yielding $\tau = 0/0$, which happens exactly when the nth rows of A and B are proportional. But again this is not really a failure at all, since it allows $\hat{a}_{nn}/\hat{b}_{nn}$ to be extracted as an eigenvalue and the problem to be deflated to a smaller one.

This discussion helps motivate the following definition. A Hessenberg pair is called a *proper Hessenberg pair* if three conditions hold: (i) $|a_{j+1,j}| + |b_{j+1,j}| > 0$ for $j = 1, \ldots, n-1$, (ii) the first columns of A and B are not proportional, (iii) the last rows of A and B are not proportional. The first condition just says that for each j, at least one of $a_{j+1,j}$ and $b_{j+1,j}$ is nonzero. If this condition is not satisfied, we can immediately reduce the pencil to two smaller pencils. If either of conditions (ii) and (iii) is not satisfied, we can also reduce the problem, as we know from the discussion immediately above. Therefore, we can always assume, without loss of generality, that we are working with a proper Hessenberg pair.

Proposition 4.1.2. *[25] In a proper Hessenberg pair, the core transformation Q_1 that replaces pole σ_1 by ρ satisfies*

$$Q_1 e_1 = \delta (A - \rho B)(A - \sigma_1 B)^{-1} e_1$$

for some nonzero δ.

Proof. From our construction we have $Q_1 e_1 = \gamma^{-1}(A - \rho B)e_1$. Since σ_1 is the first pole of the pair (A, B), we have $(A - \sigma_1 B)e_1 = \check{\gamma} e_1$ for some $\check{\gamma}$. The properness assumption guarantees that both γ and $\check{\gamma}$ are nonzero. Therefore $Q_1 e_1 = \delta(A - \rho B)(A - \sigma_1 B)^{-1} e_1$, where $\delta = (\gamma \check{\gamma})^{-1}$. □

Remark 4.1.3. *The insertion of the extra factor $(A - \sigma_1 B)^{-1}$ may seem mysterious. As we shall see later, this is just what is needed for a consistent convergence theory. In the product $(A - \rho B)(A - \sigma_1 B)^{-1}$, the factor $A - \rho B$ signals that the pole ρ is entering the pencil, while the factor $(A - \sigma_1 B)^{-1}$ signals that the pole σ_1 is leaving.*

4.1. Operations on Hessenberg pairs

Proposition 4.1.4. *[25] In a proper Hessenberg pair, the core transformation Z_{n-1} that replaces pole σ_{n-1} by τ satisfies*

$$e_n^T Z_{n-1}^* = \delta\, e_n^T (A - \sigma_{n-1} B)^{-1}(A - \tau B)$$

for some nonzero δ.

The proof is left as an exercise for the reader.

The arithmetic cost of a move of type I is just the cost of multiplying A and B by a single core transformation, Q_1^* or Z_{n-1}. If the cores are Givens rotations applied in the conventional way, the cost is about $8n$ multiplications and $4n$ additions, or $12n$ (complex) flops. Different implementations could yield slightly different flop counts, but regardless of the details the cost will be $O(n)$.

Standard backward error analysis [72] shows that moves of type I are backward stable.

Interchanging two poles. (Type II move)

The second of the two allowed operations is to interchange two adjacent poles by a unitary equivalence $\hat{A} - \lambda \hat{B} = Q_j^*(A - \lambda B)Z_{j-1}$. To understand this, consider the pole pencil $A_\pi - \lambda B_\pi$ obtained by discarding the first row and last column from $A - \lambda B$. This pencil is upper triangular and has $\sigma_1, \ldots, \sigma_{n-1}$ as its eigenvalues. Therefore the process of swapping two poles of $A - \lambda B$ is exactly that of interchanging two eigenvalues of the pole pencil $A_\pi - \lambda B_\pi$, a task that we discussed in detail in Section 2.2. There we described a process to interchange any two eigenvalues σ_{j-1} and σ_j of a triangular pencil. The exchange never fails and is backward stable.

This requires only an equivalence transformation $\tilde{Q}_{j-1}^*(A_\pi - \lambda B_\pi)\tilde{Z}_{j-1}$ by two core transformations $\tilde{Q}_{j-1}$ and $\tilde{Z}_{j-1}$ of dimension $n-1$. We then enlarge these matrices by adjoining a row and column to the top of $\tilde{Q}_{j-1}$ and the bottom of $\tilde{Z}_{j-1}$:

$$Q_j = \begin{bmatrix} 1 & 0 \\ 0 & \tilde{Q}_{j-1} \end{bmatrix}, \qquad Z_{j-1} = \begin{bmatrix} \tilde{Z}_{j-1} & 0 \\ 0 & 1 \end{bmatrix}.$$

Then $\hat{A} - \lambda \hat{B} = Q_j^*(A - \lambda B)Z_{j-1}$ is the desired transformation.

The flop count for a move of type II is about the same as for a move of type I, namely $12n$ if the core transformations are implemented as Givens rotations. In any event, the flop counts for moves of type I and II are about the same, and each move costs $O(n)$ flops.[9]

Exercises 4.1

1. Prove Proposition 4.1.4.

2. We have one type of move that is able to change a pole at one end or the other and another type that swaps poles in the middle. It is natural to ask whether we can devise a move that changes a single pole in the middle. Show that this is impossible. Specifically, consider a transformation

$$\hat{A} - \lambda \hat{B} = Q^*(A - \lambda B)Z, \tag{4.1.2}$$

where Q does not touch the first row and Z does not touch the last column. That is,

$$Q = \begin{bmatrix} 1 & \\ & \tilde{Q} \end{bmatrix} \quad \text{and} \quad Z = \begin{bmatrix} \tilde{Z} & \\ & 1 \end{bmatrix}.$$

---

[9] This is the correct count for the case when only eigenvalues are being computed. If eigenvectors or some deflating subspaces are wanted as well, the transforming matrices Q and Z also need to be updated on each move. This adds about $6n$ (complex) flops for a type I move and $12n$ flops for a type II, but the total is still $O(n)$.

Show that such a transformation cannot change the poles. (It can move them around, but it cannot introduce anything new. Thus any transformation meant to change a pole must touch either the first row or the last column. That's what the moves of type I do.)

4.2 ▪ Building an algorithm from the pieces

Suppose we want to find the eigenvalues of some regular pair (A, B). As usual, there are two steps to the process. The first is a direct method that transforms (A, B) to a condensed form, in our case a Hessenberg pencil. The second step is an iterative process that uncovers the eigenvalues of the condensed form.

Of course we can often skip the reduction phase. Notably, the rational Krylov process, which we used as motivation for this chapter, produces a pencil that is already Hessenberg.

Reduction to a Hessenberg pencil

If a reduction is required, we can do it. Moler and Stewart [50] showed how to reduce (A, B) to Hessenberg-triangular form by a direct method in $O(n^3)$ flops. The reduction is also described in [30, 69, 70] and elsewhere. Blocked versions leveraging modern computer architectures are available [36, 26]. If the resulting pair is not proper, we can split it into smaller proper pairs, so let us assume it is proper. This is a Hessenberg pencil with all poles equal to ∞. If the user is happy to start from this configuration, s/he can move directly to the iterative phase.

If the user wants to set certain prescribed poles $\sigma_1, \ldots, \sigma_{n-1}$ before beginning the iterations, this is also possible. One obvious procedure is to begin by introducing σ_{n-1} at the top of the pencil by a move of type I. Then σ_{n-1} can be swapped with each of the remaining infinite poles by moves of type II until it arrives at its desired position at the bottom. The total number of moves is $n - 1$. Then σ_{n-2} can be introduced at the top by a move of type I. It can then be swapped with each of the remaining infinite poles until it arrives at its desired position just above σ_{n-1}. The total number of moves for this step is $n - 2$. Then σ_{n-3} can be introduced, and so on. Eventually we get each of $\sigma_1, \ldots, \sigma_{n-1}$ into its desired position. The total number of moves for this phase is about $n^2/2$, and the total flop count is $O(n^3)$.

One can equally well introduce the poles at the bottom and swap them upward, starting with σ_1, then σ_2, and so on. The amount of work is exactly the same, about $n^2/2$ moves. Better yet, one can take $k \approx (n - 1)/2$ and introduce $\sigma_1, \ldots, \sigma_k$ (in reverse order) at the top and σ_{k+1}, ..., σ_{n-1} at the bottom. This cuts the number of moves in half. However one does it, the cost is $O(n^3)$.

Camps, Meerbergen, and Vandebril [25] describe a procedure that introduces the poles during the reduction to Hessenberg form. They also present an example where a good choice of poles induces a deflation in the middle of the pencil.

The iterative phase (basic algorithm)

During the discussion of moves of type I in Section 4.1 we defined *proper* Hessenberg pairs and noted that if a Hessenberg pair is not proper, it can be reduced to smaller pairs that are. We therefore assume, without loss of generality, that we have a proper Hessenberg pair (A, B) with poles $\sigma_1, \ldots, \sigma_{n-1}$. We now describe an iteration of the RQZ algorithm proposed in [25]. We will call this the *basic algorithm*.

First a shift ρ is chosen. Any of the usual shifting strategies can be employed here. The simplest is the Rayleigh-quotient shift $\rho = a_{nn}/b_{nn}$. Then ρ is introduced as a pole at the top of the pencil, replacing σ_1, by a move of type I. Next ρ is swapped with σ_2 by a move of type II. Then another move of type II is used to swap ρ with σ_3, and so on. After $n - 2$ moves of

4.2. Building an algorithm from the pieces

type II, ρ arrives at the bottom of the pencil. The poles are now $\sigma_2, \ldots, \sigma_{n-1}$, and ρ. Finally a move of type I is used to remove the pole ρ from the bottom, replacing it by a new pole σ_n. This completes the iteration. The user has complete flexibility in the choice of σ_n. One possibility is $\sigma_n = \infty$. Another, which might be called a *Rayleigh-quotient pole*, is $\sigma_n = a_{11}/b_{11}$.

The cost of one iteration of the basic algorithm is n moves or $O(n^2)$ flops. With any of the standard shifting strategies, e.g., Rayleigh-quotient shift, repeated iterations will normally cause rapid convergence of an eigenvalue at the bottom of the pencil. Typically $a_{n,n-1} \to 0$ and $b_{n,n-1} \to 0$ quadratically, leaving a_{nn}/b_{nn} as an eigenvalue and allowing deflation of the problem. After $n-1$ deflations, all of the eigenvalues will have been found.

There are numerous variations on the basic algorithm. For example, it can be turned upside down. We can pick a shift, say $\rho = a_{11}/b_{11}$, insert it at the bottom of the pencil, and chase it to the top. Since we can do this, then why not chase shifts in both directions at once? Some possibilities along these lines will be discussed in Section 4.4.

Relationship to the QZ algorithm

We now show that when the basic algorithm is applied to a pair that has all poles equal to infinity, it reduces to the single-shift version of the Moler/Stewart QZ algorithm. Consider a Hessenberg-triangular pair

$$\begin{bmatrix} \times & \times & \times & \times \\ \times & \times & \times & \times \\ & \times & \times & \times \\ & & \times & \times \end{bmatrix} \quad \begin{bmatrix} \times & \times & \times & \times \\ & \times & \times & \times \\ & & \times & \times \\ & & & \times \end{bmatrix},$$

which has poles ∞, ∞, and ∞. An iteration of the basic algorithm begins by choosing a shift ρ and inserting it into the pair at the top by a move of type I. The transformation is $A \to Q_1^* A$, $B \to Q_1^* B$, where Q_1 satisfies (4.1.1). This is exactly the same as the transformation that starts single-shift QZ [70, p. 537]. It alters the first two rows of the matrices, so the transformed matrices have the form

$$\begin{bmatrix} \times & \times & \times & \times \\ \times & \times & \times & \times \\ & \times & \times & \times \\ & & \times & \times \end{bmatrix} \quad \begin{bmatrix} \times & \times & \times & \times \\ + & \times & \times & \times \\ & & \times & \times \\ & & & \times \end{bmatrix}. \tag{4.2.1}$$

The triangular form of B has been disturbed, but this is still a Hessenberg pair. Its poles are ρ, ∞, ∞. (We will continue to refer to the matrices as "A" and "B", even though they change in the course of the iteration.) The next step of the basic algorithm is a move of type II that interchanges the pole ρ with the adjacent pole ∞, resulting in

$$\begin{bmatrix} \times & \times & \times & \times \\ \times & \times & \times & \times \\ & \times & \times & \times \\ & & \times & \times \end{bmatrix} \quad \begin{bmatrix} \times & \times & \times & \times \\ & \times & \times & \times \\ & + & \times & \times \\ & & & \times \end{bmatrix}, \tag{4.2.2}$$

a Hessenberg pair with poles ∞, ρ, ∞. The transformation has the form $A \to Q_2^* A Z_1$, $B \to Q_2^* B Z_1$, with appropriately chosen core transformations Z_1 and Q_2. Let us consider now how things look if we apply the cores one at a time. Starting from the configuration shown in (4.2.1), first apply Z_1 on the right. This acts on columns one and two of each matrix and produces

$$\begin{bmatrix} \times & \times & \times & \times \\ \times & \times & \times & \times \\ + & \times & \times & \times \\ & & \times & \times \end{bmatrix} \quad \begin{bmatrix} \times & \times & \times & \times \\ & \times & \times & \times \\ & & \times & \times \\ & & & \times \end{bmatrix}.$$

The entry b_{21} must now be zero. This is so because, as we know, after the application of Q_2^* on the left, b_{21} must be zero, as shown in (4.2.2). The left multiplication by Q_2^* cannot do this job, so it must have been done by Z_1. At the same time, Z_1 must produce a bulge at a_{31}. This proves that Z_1 is exactly the same transformation as is used at this point in the QZ bulge chase.

Now, when we apply Q_2^* on the left, it operates on rows two and three. It must set a_{31} to zero and create a new bulge at b_{32} to arrive at (4.2.2). Thus Q_2 is exactly the same transformation as is used at this point in the QZ bulge chase.

The next step is a move of type II that transforms (4.2.2) to

$$\begin{bmatrix} \times & \times & \times & \times \\ \times & \times & \times & \times \\ & \times & \times & \times \\ & & \times & \times \end{bmatrix} \quad \begin{bmatrix} \times & \times & \times & \times \\ & \times & \times & \times \\ & & \times & \times \\ & & + & \times \end{bmatrix}, \qquad (4.2.3)$$

a Hessenberg pair with poles ∞, ∞, ρ. The transformation has the form $A \to Q_3^* A Z_2$, $B \to Q_3^* B Z_2$. Again we could look at what happens if we apply the cores one at a time, first Z_2, then Q_3^*, and we would find as before that these are exactly the same transformations as in a QZ bulge chase.

In our little example, we have now reached the bottom. In a larger example, we would continue moves of type II, pushing the pole ρ downward, and at each step we would have the same situation. The final step is a move of type I that removes ρ from the bottom of the pencil, replacing it by a pole ∞. This is exactly the transform, acting on columns $n-1$ and n, that sets $b_{n,n-1}$ (the b_{43} entry in (4.2.3)) to zero. Again this is exactly the same as the transformation that completes the QZ bulge chase. The pair is now in Hessenberg-triangular form.

We have demonstrated that the basic algorithm reduces to the single-shift QZ algorithm in the case when all of the poles are infinite.

Exercises 4.2

1. In our discussion of the reduction to Hessenberg form, we showed how to insert finite poles if desired. Confirm that the approximate move counts stated there, $n^2/2$ and $n^2/4$, are correct.

4.3 ▪ Convergence theory

In our convergence theorems we make the blanket (and generically valid) assumption that none of the poles or shifts that are mentioned are eigenvalues of the pencil.

The mechanism that drives all variants of Francis's algorithm is nested subspace iteration with changes of coordinate system [70, p. 431], [71, p. 399], [3, Thm. 2.2.3]. As a specific example, let us consider a single step of the QZ algorithm with shift ρ applied to a Hessenberg-triangular pencil $A - \lambda B$, yielding a new pencil $\hat{A} - \lambda \hat{B}$ with

$$\hat{A} - \lambda \hat{B} = Q^*(A - \lambda B)Z. \qquad (4.3.1)$$

First we define some nested sequences of subspaces. For $k = 1, \ldots, n$, define

$$\mathcal{E}_k = \text{span}\{e_1, \ldots, e_k\},$$

where $e_1, \ldots, e_n$ are the standard basis vectors. Then define

$$\mathcal{Q}_k = Q\mathcal{E}_k \quad \text{and} \quad \mathcal{Z}_k = Z\mathcal{E}_k.$$

Thus $\mathcal{Q}_k$ (resp., $\mathcal{Z}_k$) is the space spanned by the first k columns of Q (resp., Z).

4.3. Convergence theory

Theorem 4.3.1. *A single step of the QZ algorithm with shift ρ effects nested subspace iterations*

$$\mathcal{Q}_k = (AB^{-1} - \rho I)\mathcal{E}_k, \quad \mathcal{Z}_k = (B^{-1}A - \rho I)\mathcal{E}_k, \quad k = 1, \ldots, n-1.$$

The change of coordinate system (4.3.1) transforms both $\mathcal{Q}_k$ and $\mathcal{Z}_k$ back to $\mathcal{E}_k$.

We call this a *convergence theorem* even though it makes no mention of convergence. Theorems like this can be used together with the convergence theory of subspace iteration to draw conclusions about the convergence of the algorithm, as explained in [69, 70, 71] and elsewhere.

Camps, Meerbergen, and Vandebril [25, Thm. 6.1] proved a result like Theorem 4.3.1 for the basic algorithm. The scenario is similar. The iteration begins with a proper Hessenberg pair (A, B) with poles $\sigma_1, \ldots, \sigma_{n-1}$, employs a shift ρ, and ends with a new proper Hessenberg pair $(\hat{A}, \hat{B})$ with poles $\sigma_2, \ldots, \sigma_n$. The old and new pairs are related by a unitary equivalence transformation of the form (4.3.1).

Theorem 4.3.2. *A single step of the basic algorithm with shift ρ, starting with a proper Hessenberg pair (A, B) with poles $\sigma_1, \ldots, \sigma_{n-1}$ and ending with $(\hat{A}, \hat{B})$ with poles $\sigma_2, \ldots, \sigma_n$ effects nested subspace iterations*

$$\mathcal{Q}_k = (A - \rho B)(A - \sigma_k B)^{-1}\mathcal{E}_k, \quad \mathcal{Z}_k = (A - \sigma_{k+1}B)^{-1}(A - \rho B)\mathcal{E}_k, \quad k = 1, \ldots, n-1.$$

The change of coordinate system (4.3.1) transforms both $\mathcal{Q}_k$ and $\mathcal{Z}_k$ back to $\mathcal{E}_k$.

This theorem was proved in [25], but we will also provide a proof based on our new theory. Comparing this with Theorem 4.3.1, we see that the inclusion of poles gives extra freedom that might be used to improve convergence.

Now consider Theorem 4.3.2 in the case when all of the poles are infinite. When $\sigma_k = \infty$, the operator $(A - \rho B)(A - \sigma_k B)^{-1}$ becomes (when appropriately rescaled) $(A - \rho B)B^{-1} = AB^{-1} - \rho I$. Similarly $(A - \sigma_{k+1}B)^{-1}(A - \rho B)$ becomes $B^{-1}(A - \rho B) = B^{-1}A - \rho I$. These operators are exactly the ones that appear in Theorem 4.3.1, just as we would expect.

Although the QZ algorithm is a special case of the basic algorithm, there is an important difference in the way they are viewed. The QZ algorithm acts on proper Hessenberg-triangular pencils. It is a bulge-chasing algorithm. The initial equivalence transformation of each iteration creates a bulge in the Hessenberg-triangular form. The rest of the iteration consists of equivalence transformations that chase the bulge back and forth between A and B until it finally disappears off of the bottom of the pencil. At that point the Hessenberg-triangular form has been restored and the iteration is complete. The QZ algorithm can also be implemented as a *core-chasing* algorithm, as is shown in [3] and [5], but the situation is the same: The Hessenberg-triangular form is disturbed at the beginning of the iteration and not restored until the very end.

Now let us contrast this with what happens in the basic algorithm (with infinite poles or otherwise). The basic algorithm operates on proper Hessenberg pairs, in which neither matrix is required to be triangular. Each iteration starts with a move of type I, performs a sequence of moves of type II, and ends with a move of type I. These moves do not disturb the Hessenberg form; it is preserved throughout. This implies that we can think of each move as a "mini-iteration" and ask whether we can obtain a result like Theorem 4.3.1 or 4.3.2 for each individual move of type I or II. It turns out that we can.

Each move of either type is an equivalence transform of the form

$$\hat{A} = Q_j^* A Z_{j-1}, \quad \hat{B} = Q_j^* B Z_{j-1}.$$

The case $j = 1$ denotes a move of type I, and we have $Z_0 = I$. The case $j = n$ also denotes a type I move, and in this case $Q_n = I$. The cases $j = 2, \ldots, n-1$ are of type II. Suppose (A, B)

has poles $\sigma_1, \ldots, \sigma_{n-1}$. A move of type II interchanges poles σ_{j-1} and σ_j. For the moves of type I, in the case $j = 1$, suppose the pole σ_1 is replaced by a new pole σ_0; in the case $j = n$, suppose σ_{n-1} is replaced by a new pole σ_n. With this notation we can cover both types of move by a single theorem.

As above we define sequences of nested subspaces $(\mathcal{Q}_k)$ and $(\mathcal{Z}_k)$, where $\mathcal{Q}_k$ (resp., $\mathcal{Z}_k$) is the space spanned by the first k columns of Q_j (resp., Z_{j-1}). But note that, because Q_j and Z_{j-1} are core transformations, these spaces are mostly trivial in this setting: $\mathcal{Q}_k = \mathcal{E}_k$ except when $k = j$, and $\mathcal{Z}_k = \mathcal{E}_k$ except when $k = j-1$.

Theorem 4.3.3. *Using notation and terminology established directly above, the move*

$$\hat{A} - \lambda \hat{B} = Q_j^*(A - \lambda B)Z_{j-1} \tag{4.3.2}$$

effects nested subspace iterations that are, however, mostly trivial. The nontrivial actions are

$$\mathcal{Q}_j = (A - \sigma_{j-1}B)(A - \sigma_j B)^{-1}\mathcal{E}_j$$

and

$$\mathcal{Z}_{j-1} = (A - \sigma_j B)^{-1}(A - \sigma_{j-1}B)\mathcal{E}_{j-1}.$$

The change of coordinate system (4.3.2) transforms $\mathcal{Q}_j$ back to $\mathcal{E}_j$ and $\mathcal{Z}_{j-1}$ back to $\mathcal{E}_{j-1}$.

The proof, which requires some preparation, is deferred to Section 4.7.

Using Theorem 4.3.3

In all of our convergence theorems we have actions of the form $\mathcal{Q}_k = r(AB^{-1})\mathcal{E}_k$ and $\mathcal{Z}_k = r(B^{-1}A)\mathcal{E}_k$, where r is a rational function, e.g., $r(z) = (z - \sigma_{j-1})/(z - \sigma_j)$. In the following lemma the functions r and s can be any functions defined on the spectrum of the pencil $A - \lambda B$, but in our applications they will always be rational. In this case, being defined on the spectrum of $A - \lambda B$ just means that none of the poles is an eigenvalue.

Lemma 4.3.4. *Consider two successive changes of coordinate system*

$$\tilde{A} - \lambda \tilde{B} = \tilde{Q}^*(A - \lambda B)\tilde{Z} \quad \text{and} \quad \hat{A} - \lambda \hat{B} = \hat{Q}^*(\tilde{A} - \lambda \tilde{B})\hat{Z},$$

so that

$$\hat{A} - \lambda \hat{B} = Q^*(A - \lambda B)Z, \quad \text{where} \quad Q = \tilde{Q}\hat{Q} \quad \text{and} \quad Z = \tilde{Z}\hat{Z}.$$

For $k = 1, \ldots, n-1$, if

$$\tilde{Q}\mathcal{E}_k = r(AB^{-1})\mathcal{E}_k \quad \text{and} \quad \hat{Q}\mathcal{E}_k = s(\tilde{A}\tilde{B}^{-1})\mathcal{E}_k,$$

then

$$Q\mathcal{E}_k = sr(AB^{-1})\mathcal{E}_k,$$

where sr is the pointwise product of s and r. If

$$\tilde{Z}\mathcal{E}_k = r(B^{-1}A)\mathcal{E}_k \quad \text{and} \quad \hat{Z}\mathcal{E}_k = s(\tilde{B}^{-1}\tilde{A})\mathcal{E}_k,$$

then

$$Z\mathcal{E}_k = sr(B^{-1}A)\mathcal{E}_k.$$

4.3. Convergence theory

Proof. Noting that $\tilde{Q}\, s(\tilde{A}\tilde{B}^{-1}) = s(AB^{-1})\tilde{Q}$, we have

$$Q\mathcal{E}_k = \tilde{Q}\hat{Q}\mathcal{E}_k = \tilde{Q}\, s(\tilde{A}\tilde{B}^{-1})\mathcal{E}_k = s(AB^{-1})\tilde{Q}\mathcal{E}_k = s(AB^{-1})r(AB^{-1})\mathcal{E}_k,$$

so $Q\mathcal{E}_k = sr(AB^{-1})\mathcal{E}_k$. The result for $Z\mathcal{E}_k$ is proved similarly, using the equation $\tilde{Z}\, s(\tilde{B}^{-1}\tilde{A}) = s(B^{-1}A)\tilde{Z}$. □

Clearly this lemma can be extended by induction to three or more successive changes of coordinate system, and that's how we are going to use it.

Proof of Theorem 4.3.2

As a first application of Theorem 4.3.3, we show that it can be used to prove Theorem 4.3.2.

According to Theorem 4.3.2, for each k the basic algorithm effects a transformation

$$Q_k = (A - \rho B)(A - \sigma_k B)^{-1}\mathcal{E}_k. \tag{4.3.3}$$

Let us see why this is so. Recall that the basic algorithm begins with a move of type I that introduces the shift ρ as a pole at the top of the pencil. It then does a sequence of moves of type II that swap ρ with the other poles one by one. For a given k, most of these moves have no effect on $\mathcal{E}_k$. The only exception is the kth move, the case $j = k$ in Theorem 4.3.3. This is where we need to focus.

One iteration of the basic algorithm performs the equivalence

$$\hat{A} - \lambda\hat{B} = Q^*(A - \lambda B)Z,$$

where Q and Z are products of core transformations:

$$Q = Q_1 Q_2 \cdots Q_{n-1}, \qquad Z = Z_1 Z_2 \cdots Z_{n-1}.$$

The core Q_1 is the one that replaces pole σ_1 with the shift ρ. Q_2 (together with Z_1) swaps ρ with σ_2, Q_3 (together with Z_2) swaps ρ with σ_3, and so on. Z_{n-1} removes ρ and installs a new pole σ_n. We are interested in the action of Q_k (together with Z_{k-1}), which swaps ρ with σ_k. Thus we factor Q and Z as

$$Q = \tilde{Q} Q_k \hat{Q}, \qquad Z = \tilde{Z} Z_{k-1} \hat{Z},$$

where $\tilde{Q} = Q_1 \cdots Q_{k-1}$, and so on. Now we break the transformation into three parts:

$$\tilde{A} - \lambda\tilde{B} = \tilde{Q}^*(A - \lambda B)\tilde{Z},$$

$$\check{A} - \lambda\check{B} = Q_k^*(\tilde{A} - \lambda\tilde{B})Z_{k-1}, \tag{4.3.4}$$

and

$$\hat{A} - \lambda\hat{B} = \hat{Q}^*(\check{A} - \lambda\check{B})\hat{Z}.$$

Because each of the cores $Q_1, \ldots, Q_{k-1}$ leaves $\mathcal{E}_k$ invariant, we have

$$\tilde{Q}\mathcal{E}_k = \mathcal{E}_k = r(AB^{-1})\mathcal{E}_k, \quad \text{where } r(z) = 1.$$

We can apply Theorem 4.3.3 with $j = k$ to the transformation (4.3.4), taking into account that the poles that are swapped in the kth move are ρ and σ_k, to get

$$Q_k\mathcal{E}_k = (\tilde{A} - \rho\tilde{B})(\tilde{A} - \sigma_k\tilde{B})^{-1}\mathcal{E}_k = s(\tilde{A}\tilde{B}^{-1})\mathcal{E}_k, \quad \text{where } s(z) = (z - \rho)/(z - \sigma_k).$$

Finally, noting that $Q_{k+1}, \ldots, Q_{n-1}$ all leave $\mathcal{E}_k$ invariant, we have
$$\hat{Q}\mathcal{E}_k = \mathcal{E}_k = t(\check{A}\check{B}^{-1})\mathcal{E}_k, \quad \text{where } t(z) = 1.$$
Now, applying Lemma 4.3.4 to the product $Q = \tilde{Q}Q_k\hat{Q}$, we get
$$Q\mathcal{E}_k = tsr(AB^{-1})\mathcal{E}_k = s(AB^{-1})\mathcal{E}_k = (A - \rho B)(A - \sigma_k B)^{-1}\mathcal{E}_k,$$
which is exactly (4.3.3).

We can prove the Z part of Theorem 4.3.2 in exactly the same way. We have
$$\tilde{Z}\mathcal{E}_{k-1} = \mathcal{E}_{k-1} = 1(B^{-1}A)\mathcal{E}_{k-1},$$
and by Theorem 4.3.3 with $j = k$,
$$Z_{k-1}\mathcal{E}_{k-1} = (\tilde{A} - \sigma_k\tilde{B})^{-1}(\tilde{A} - \rho\tilde{B})\mathcal{E}_{k-1} = s(\tilde{B}^{-1}\tilde{A})\mathcal{E}_{k-1},$$
and finally
$$\hat{Z}\mathcal{E}_{k-1} = \mathcal{E}_{k-1} = 1(\check{B}^{-1}\check{A})\mathcal{E}_{k-1}.$$
Therefore, by Lemma 4.3.4,
$$Z\mathcal{E}_{k-1} = s(B^{-1}A)\mathcal{E}_{k-1} = (A - \sigma_k B)^{-1}(A - \rho B)\mathcal{E}_{k-1}.$$
Adding one to the index k, we get the Z part of Theorem 4.3.2, thereby completing the proof.

Generalization of the proof

The basic algorithm is just one of many possible algorithms that make use of moves of types I and II on proper Hessenberg forms. We have already pointed out that one could run the algorithm in the opposite direction or in both directions at once. There are lots of other possibilities, and we will look at some in what follows.

From our proof of Theorem 4.3.2 it should now be clear that we will be able to use Theorem 4.3.3, together with Lemma 4.3.4, to analyze the action of any algorithm that acts on a proper Hessenberg pencil by moves of types I and II. Consider a transformation
$$\hat{A} - \lambda\hat{B} = Q^*(A - \lambda B)Z, \tag{4.3.5}$$
where Q and Z are products of core transformations generated by any sequence of moves of type I and II. If we want to find the action of Q on $\mathcal{E}_k$ for some k, we need only look at the core transformations of the form Q_k, i.e., the ones that act in the $(k, k+1)$ plane. Thus we factor Q into a product of the form
$$Q = \tilde{Q}Q_{1,k}\check{Q}Q_{2,k}\hat{Q}Q_{3,k}\cdots, \tag{4.3.6}$$
where $\tilde{Q}, \check{Q}, \ldots$ are products of core transformations that do not act in the $(k, k+1)$ plane and therefore satisfy $\tilde{Q}\mathcal{E}_k = \mathcal{E}_k$, $\check{Q}\mathcal{E}_k = \mathcal{E}_k$, and so on, and $Q_{1,k}, Q_{2,k}, \ldots$ are cores that do act in the $(k, k+1)$ plane. Let us say there are m such cores $Q_{1,k}, \ldots, Q_{m,k}$.

The transforming matrix Z has a fully analogous factorization
$$Z = \tilde{Z}Z_{1,k-1}\check{Z}Z_{2,k-1}\hat{Z}Z_{3,k-1}\cdots. \tag{4.3.7}$$
We have $\tilde{Z}\mathcal{E}_{k-1} = \mathcal{E}_{k-1}$, $\check{Z}\mathcal{E}_{k-1} = \mathcal{E}_{k-1}$, etc. The transformations that act nontrivially on $\mathcal{E}_{k-1}$ are $Z_{1,k-1}, \ldots, Z_{m,k-1}$.

4.4. Variations on the basic algorithm

Suppose that on the move corresponding to the transformations $Q_{j,k}$ and $Z_{j,k-1}$, the poles that get swapped are $\sigma_{j,k-1}$ and $\sigma_{j,k}$. Then, according to Theorem 4.3.3, the function associated with this swap is $r_j(z) = (z - \sigma_{j,k-1})/(z - \sigma_{j,k})$. Let r denote the product of these functions:

$$r(z) = r_1(z) \cdots r_m(z) = \prod_{j=1}^{m} \frac{z - \sigma_{j,k-1}}{z - \sigma_{j,k}}. \qquad (4.3.8)$$

Then, applying Lemma 4.3.4 to the long product of transformations defined by (4.3.6) and (4.3.7), we find that the action of Q on $\mathcal{E}_k$ and of Z on $\mathcal{E}_{k-1}$ is given by

$$\mathcal{Q}_k = Q\mathcal{E}_k = r(AB^{-1})\mathcal{E}_k \quad \text{and} \quad \mathcal{Z}_{k-1} = Z\mathcal{E}_{k-1} = r(B^{-1}A)\mathcal{E}_{k-1}. \qquad (4.3.9)$$

We summarize these findings as a theorem.

Theorem 4.3.5. *Consider a transformation (4.3.5), where Q and Z are products of core transformations generated by any sequence of moves of types I and II. For some k suppose that m of the moves acted at the kth position, swapping poles $\sigma_{j,k-1}$ and $\sigma_{j,k}$ for $j = 1, \ldots, m$. Define a rational function r by (4.3.8). Then the action of Q on $\mathcal{E}_k$ and of Z on $\mathcal{E}_{k-1}$ is given by (4.3.9). The transformation (4.3.5) transforms $\mathcal{Q}_k$ back to $\mathcal{E}_k$ and $\mathcal{Z}_{k-1}$ back to $\mathcal{E}_{k-1}$.*

Exercises 4.3

1. Show that in the move described in Theorem 4.3.3, we have $\mathcal{Q}_k = \mathcal{E}_k$ if $k \neq j$ and $\mathcal{Z}_k = \mathcal{E}_k$ if $k \neq j - 1$.

4.4 ▪ Variations on the basic algorithm

In this section we consider algorithms built exclusively from moves of types I and II. Since the moves are backward stable, the resulting algorithms are also backward stable.

The basic algorithm (like the single-shift bulge-chasing and core-chasing algorithms) takes a single shift, inserts it into the top of the pencil, and chases it to the bottom. This algorithm suffers from inefficient use of cache memory and negligible potential for parallelism. In the case of bulge-chasing algorithms the problem was remedied by selecting a large number of shifts at once, creating many small bulges one after the other, and chasing this chain of bulges together to the bottom of the matrix or pencil [16, 45, 46]. This allows the use of Level 3 BLAS and therefore efficient cache use. It also provides an opportunity for parallelism [33].

Chasing multiple shifts at once

The same remedy works for pole-swapping algorithms, as was already mentioned in [20, 25, 58]. We can choose m shifts $\rho_1, \ldots, \rho_m$, where typically $1 \ll m \ll n$.[10] Suppose the poles of $A - \lambda B$ are

$$\sigma_1, \ldots, \sigma_m, \sigma_{m+1}, \ldots, \sigma_n.$$

By a sequence of moves of types I and II we can replace $\sigma_1, \ldots, \sigma_m$ by $\rho_1, \ldots, \rho_m$, so that the poles of the new pencil are

$$\rho_1, \ldots, \rho_m, \sigma_{m+1}, \ldots, \sigma_n.$$

Then we can chase these m shifts together to the bottom, creating enough arithmetic to make efficient use of cache. To be precise, in the first step we would swap σ_{m+1} with ρ_m, then σ_{m+1}

---

[10]One way to obtain m shifts is to use an auxiliary routine to compute the eigenvalues of the lower-right-hand $m \times m$ subpencil of $A - \lambda B$, and use these as the shifts.

with ρ_{m-1}, and so on. Eventually we swap σ_{m+1} with ρ_1, putting σ_{m+1} at the top. Then we go on to the next step.

We can pass a chain of shifts from top to bottom, and we can equally well pass a chain from bottom to top. If we wish, we can pass chains in both directions at once. Suppose we have shifts $\rho_1, \ldots, \rho_m$ that we wish to chase from top to bottom and shifts $\tau_1, \ldots, \tau_m$ that we wish to chase from bottom to top. Using moves of types I and II we can introduce them:

$$\rho_1, \ldots, \rho_m, \sigma_{m+1}, \ldots, \sigma_{n-m-1}, \tau_1, \ldots, \tau_m.$$

We then chase the ρ's downward and the τ's upward. The two chains pass through each other,[11] and eventually we get to the position

$$\tau_1, \ldots, \tau_m, \sigma_{m+1}, \ldots, \sigma_{n-m-1}, \rho_1, \ldots, \rho_m.$$

The reader can check that the poles in the middle, $\sigma_{m+1}, \ldots, \sigma_{n-m-1}$, get moved around in the process, but they end up exactly where they started. At this point we can regard the iteration as complete, or we can "complete" the iteration by removing the τ_i and ρ_i from the pencil and replacing them with new sets of shifts.

Let's see what Theorem 4.3.5 tells us about this bidirectional procedure. Let

$$r(z) = \prod_{i=1}^{m} \frac{z - \rho_i}{z - \tau_i}. \tag{4.4.1}$$

Then for $k = m+1, \ldots, n-m$ we have the action

$$\mathcal{Q}_k = Q\mathcal{E}_k = r(AB^{-1})\mathcal{E}_k \quad \text{and} \quad \mathcal{Z}_{k-1} = Z\mathcal{E}_{k-1} = r(B^{-1}A)\mathcal{E}_{k-1}.$$

The reason for this is that each of the ρ_i passes downward through the kth position, causing a factor $z - \rho_i$, and each of the τ_i passes upward, causing a factor $(z - \tau_i)^{-1}$. This isn't all that happens at position k, but it's all that matters. To see this, consider, for example, a position k at which all of the ρ_i pass through before any of the τ_i get there. Passing each ρ_i downward requires also passing a σ_j upward, causing a factor $(z - \sigma_j)^{-1}$. Later on, when the τ_i are being passed upward, each σ_j that was previously passed upward gets passed downward through the kth position, causing a factor $z - \sigma_j$. The factors $(z - \sigma_j)^{-1}$ and $z - \sigma_j$ cancel each other out. We know that this must happen for each σ_j because each σ_j starts and ends in the same position.

This analysis shows that if we want to implement such a scheme, we must make sure that $\rho_1, \ldots, \rho_m$ are well separated from $\tau_1, \ldots, \tau_m$. If two shifts ρ_i and τ_j are very close together, the factors $z - \rho_i$ and $1/(z - \sigma_j)$ in (4.4.1) will nearly cancel each other out. In the most extreme case, where $\rho_1, \ldots, \rho_m$ are equal to $\tau_1, \ldots, \tau_m$, we will have $r(z) = 1$, and we will make no progress.

An optimistic scenario

Consider a situation in which we have in hand the information that we need to split the problem. Suppose we know a k (with $m+1 \leq k \leq n-m-1$) where (we think) we can split the pencil, and suppose that we have in mind an (m, m) rational function

$$r(z) = \prod_{i=1}^{m} \frac{z - \rho_i}{z - \tau_i}$$

---

[11] In the pole-swapping framework passing two chains through each other is straightforward, while in the bulge-chasing framework it is very much not; see, e.g., [65].

4.5. Aggressive early deflation

that can (nearly) split it. By this we mean that $r(AB^{-1})\mathcal{E}_k$ is (nearly) invariant under AB^{-1} and $r(B^{-1}A)\mathcal{E}_k$ is (nearly) invariant under $B^{-1}A$. If we then take the ρ_i as shifts to be passed downward and the τ_i as shifts to be passed upward, we will get both $\mathcal{Q}_k = Q\mathcal{E}_k = r(AB^{-1})\mathcal{E}_k$ and $\mathcal{Z}_k = Z\mathcal{E}_k = r(B^{-1}A)\mathcal{E}_k$. The change of variables $\hat{A} - \lambda\hat{B} = Q^*(A - \lambda B)Z$ maps both of these spaces back to $\mathcal{E}_k$. Thus $\mathcal{E}_k$ is (nearly) invariant under both $\hat{A}\hat{B}^{-1}$ and $\hat{B}^{-1}\hat{A}$, which implies that $(\mathcal{E}_k, \mathcal{E}_k)$ is (nearly) a deflating subspace for $(\hat{A}, \hat{B})$. If the pencil does not quite split apart, another step with the same (or improved?) shifts may get the job done. Notice that to achieve the desired spaces $\mathcal{Q}_k = r(AB^{-1})\mathcal{E}_k$ and $\mathcal{Z}_k = r(B^{-1}A)\mathcal{E}_k$, it is not necessary to pass the shifts all the way through the pencil. All that is needed is that $\rho_1, \ldots, \rho_m$ are pushed downward past position $k+1$ and $\tau_1, \ldots, \tau_m$ are passed upward past position k.

Of course this is a very optimistic scenario. (Where do we get these special shifts?) We include it here just to indicate what might be possible and to illustrate the use of Theorem 4.3.5.

Practical shift strategies

A more realistic plan is to take (for example) $\rho_1, \ldots, \rho_m$ to be the eigenvalues of the lower-right-hand $m \times m$ subpencil and $\tau_1, \ldots, \tau_m$ the eigenvalues of the upper-left-hand $m \times m$ subpencil, which will have the effect of causing deflations near the ends of the pencil.[12] An even better idea, assuming n is large enough, is to include *aggressive early deflation* [17], which is easy to implement in this context.

4.5 ▪ Aggressive early deflation

This powerful technique was introduced by Braman, Byers, and Mathias [17] for the single-matrix case, but it is easily adapted to matrix pencils. In order to keep the discussion as simple as possible, we will consider first the single-matrix case. Suppose, therefore, that we want to compute the eigenvalues of a matrix $A \in \mathbb{C}^{n \times n}$, where n is fairly large. Assume A is already in proper Hessenberg form, and we are going to compute its eigenvalues by some algorithm that requires a lot of shifts. Aggressive early deflation combines shift computation with deflation. If we need m shifts, we pick a number k that is significantly larger than m, for example, $k = 2m$, and compute the eigenvalues of the bottom $k \times k$ submatrix, some of which will be chosen as shifts. This is relatively inexpensive if $k \ll n$.[13] Partition A as

$$A = \begin{bmatrix} A_{11} & A_{12} \\ A_{21} & A_{22} \end{bmatrix}, \qquad A_{22} \in \mathbb{C}^{k \times k}. \tag{4.5.1}$$

Then $A_{11} \in \mathbb{C}^{j \times j}$, where $j = n - k$. We also note that A_{21} has only one nonzero entry and can be written as $A_{21} = a_{j+1,j} e_1 e_j^T$. We compute not only the eigenvalues of A_{22} but the entire Schur decomposition $T_{22} = Q_2^* A_{22} Q_2$, where T_{22} is upper triangular and Q_2 is unitary. Let Q be the $n \times n$ unitary matrix defined by $Q = \text{diag}\{I, Q_2\}$. Then

$$Q^* A Q = \begin{bmatrix} A_{11} & A_{12} Q_2 \\ Q_2^* A_{21} & T_{22} \end{bmatrix}.$$

In the lower left corner we have

$$Q_2^* A_{21} = a_{j+1,j} Q_2^* e_1 e_j^T.$$

---

[12]Notice, however, that a strategy like this should also include some provision to ensure that the upward-moving shifts are well separated from the downward-moving shifts. If some ρ_j is (nearly) equal to one of the τ_i, they will (nearly) cancel each other out.

[13]For example, we might have $n = 1000$, $m = 40$, and $k = 80$.

The vector $Q_2^* e_1$ represents a spike in the jth column of the matrix. Pictorially, the bottom $k \times (k+1)$ submatrix of $Q^* A Q$ looks like

$$\begin{bmatrix} \times & \times & \times & \cdots & \times \\ \times & & \times & \cdots & \times \\ \vdots & & & \ddots & \vdots \\ \times & & & & \times \end{bmatrix}. \tag{4.5.2}$$

The spike is shown in red. If all of the entries in the spike were zero (impossible!), then all of the eigenvalues of T_{22} would be eigenvalues of A, and we could deflate to a $j \times j$ problem. This will never happen, but we can check each of the eigenvalues of T_{22} individually and see if it qualifies as an eigenvalue of A.

We start at the bottom. If the bottom entry of the spike is small enough, we can set it to zero and proclaim that $\lambda_k = t_{kk}$ is an eigenvalue of A. This allows us to deflate the problem to size $(n-1) \times (n-1)$ and move on to the next spike entry. If, on the other hand, t_{kk} does not qualify as an eigenvalue of A, we can do $k-1$ eigenvalue swaps to move it to the top of T_{22}. This moves a different eigenvalue to the bottom, and the swaps also alter the entries of the spike. Now we check the new bottom spike entry and see if it is small enough to admit the new t_{kk} as an eigenvalue. If so, we deflate and move on. Otherwise we do more swaps to move another eigenvalue of T_{22} into the (k,k) position and check again. We continue this process until all k eigenvalues of T_{22} have been tested.

Suppose the process netted s eigenvalues. Then $k-s$ eigenvalues of T_{22} did not qualify. Of these we can pick out m to use as shifts for the next iteration. Before the iteration we must eliminate what remains of the spike. This amounts to reducing the bottom $(k-s) \times (k-s)$ submatrix to Hessenberg form.

If the deflation was so successful that it produced a large number of eigenvalues, it makes sense to do another round of aggressive early deflation immediately.

Pencil case

Now let's see how to do aggressive early deflation on a proper Hessenberg pencil $A - \lambda B$. Say we want m shifts to feed in at the top and pass down through the pencil in a tightly packed group, as described in Section 4.4. We compute the eigenvalues of the bottom $k \times k$ subpencil, where k is significantly larger than m. Partition A as in (4.5.1) and B similarly:

$$B = \begin{bmatrix} B_{11} & B_{12} \\ B_{21} & B_{22} \end{bmatrix}, \quad B_{22} \in \mathbb{C}^{k \times k}.$$

Letting $j = n - k$ we have $B_{11} \in \mathbb{C}^{j \times j}$. The submatrices A_{21} and B_{21} have only one nonzero entry each and can be written as

$$A_{21} = a_{j+1,j} e_1 e_j^T \quad \text{and} \quad B_{21} = b_{j+1,j} e_1 e_j^T.$$

We compute not only the eigenvalues but the entire generalized Schur decomposition of the pair (A_{22}, B_{22}) as well:

$$T_{22} - \lambda S_{22} = Q_2^* (A_{22} - \lambda B_{22}) Z_2,$$

where T_{22} and S_{22} are upper triangular, and Q_2 and Z_2 are unitary. Let Q and Z be the $n \times n$ unitary matrices defined by $Q = \text{diag}\{I, Q_2\}$ and $Z = \text{diag}\{I, Z_2\}$. Then

$$Q^* (A - \lambda B) Z = \begin{bmatrix} A_{11} & A_{12} Z_2 \\ Q_2^* A_{21} & T_{22} \end{bmatrix} - \lambda \begin{bmatrix} B_{11} & B_{12} Z_2 \\ Q_2^* A_{21} & S_{22} \end{bmatrix}.$$

4.5. Aggressive early deflation

In the lower left corner we have

$$Q_2^* A_{21} = a_{j+1,j} Q_2^* e_1 e_j^T \quad \text{and} \quad Q_2^* B_{21} = b_{j+1,j} Q_2^* e_1 e_j^T.$$

Thus the transformed A and B both have spikes in the jth column. Pictorially, the bottom $k \times (k+1)$ submatrix of both matrices looks like (4.5.2). Fortunately the two spikes are proportional, both multiples of $Q_2^* e_1$. Whatever we do to one spike, the exact same thing happens to the other.

Now we proceed exactly as in the single-matrix case. If the entry at the bottom of each spike is small enough to be set to zero, we do it. We declare t_{kk}/s_{kk} to be an eigenvalue and deflate. If the entries are not small enough, we do some eigenvalue swaps to move another eigenvalue into the bottom position. Then we test that eigenvalue and either deflate or swap some more eigenvalues. Once all eigenvalues of (T_{22}, S_{22}) have been tested, we are done.

Before we do the next iteration, we need to return the matrices to Hessenberg form by removing what is left of the spikes. This is particularly easy in this setting. Suppose, for example, that the remaining spike has length 4, as shown here:

$$\begin{bmatrix} \times & \times \times \times \times \\ \times & \times \times \times \\ \times & \times \times \\ \times & \times \end{bmatrix}.$$

This is the bottom 4×5 submatrix, which consists of a spike and a triangular part. We need to remove three of the spike entries to get back to Hessenberg form. The picture is the same for both A and B. We can do this by applying three core transformations on the left. The first one acts on rows 3 and 4:

$$\begin{array}{c} \\ \\ \\ \curvearrowleft \end{array} \begin{bmatrix} \times & \times \times \times \times \\ \times & \times \times \times \\ \times & \times \times \\ \times & \times \end{bmatrix} = \begin{bmatrix} \times & \times \times \times \times \\ \times & \times \times \times \\ \times & \times \times \\ & \times \times \end{bmatrix}.$$

This transforms a spike entry to zero and causes some fill-in in the triangular matrix. The same core transformation works for both A and B because the spikes are proportional. Continuing in this way we get

$$\begin{array}{c} \\ \\ \curvearrowleft \\ \curvearrowleft \\ \curvearrowleft \end{array} \begin{bmatrix} \times & \times \times \times \times \\ \times & \times \times \times \\ \times & \times \times \\ \times & \times \end{bmatrix} = \begin{bmatrix} \times & \times \times \times \times \\ & \times \times \times \\ & \times \times \times \\ & \times \times \end{bmatrix}.$$

Both A and B have been returned to Hessenberg form. Clearly this procedure can be applied to spikes of any length.

If the aggressive early deflation netted a large number of eigenvalues, we can immediately do another round of aggressive early deflation. Otherwise we pick, for use in the next iteration, m shifts from among the eigenvalues of (A_{22}, B_{22}) that did not get deflated.

We have portrayed aggressive early deflation at the bottom of the pencil, but we can also do it at the top by an analogous procedure. This especially makes sense in a situation where we are chasing large numbers of shifts in both directions, but it also makes sense if we are just chasing shifts from top to bottom. In this case we are also introducing poles at the bottom at the end of each iteration. These poles move slowly upward and, if well chosen, can eventually cause deflations at the top.

Numerical experiments investigating aggressive early deflation for pole-swapping algorithms can be found in [58].

Exercises 4.5

1. Sketch the process for aggressive early deflation at the top of the pencil.

4.6 • Connections to earlier work
Bulge pencils

The purpose of shifting is to accelerate convergence. In the standard Francis bulge-chasing algorithm the shifts are inserted at the top. That is, the shifts are used to help determine the initial transformation that creates the bulge. Then the shifts are forgotten, and the bulge is chased downward until it disappears off the bottom. Well-chosen shifts, inserted at the top, lead to rapid emergence of eigenvalues at the bottom of the matrix or pencil. Thus the information about the shifts is somehow transmitted in the bulge from top to bottom.

About thirty years ago one of the authors began to study the mechanism by which the shift information is conveyed in bulge-chasing algorithms. This study took some time, it seemed to be nontrivial, and it led to the discovery of the *bulge pencil* [66, 67, 69].

Now let's take a fresh look at the bulge pencil in light of what we now know about pole swapping. Suppose we pick a single shift ρ and begin chasing a bulge downward in a Hessenberg-triangular pair (A, B). After a couple of steps we have

$$\begin{bmatrix} \times \times \times \times \times \times \\ \times \times \times \times \times \times \\ \times \times \times \times \times \\ + \times \times \times \times \\ \times \times \times \\ \times \times \end{bmatrix} \begin{bmatrix} \times \times \times \times \times \times \\ \times \times \times \times \times \\ \times \times \times \times \\ \times \times \times \\ \times \times \\ \times \end{bmatrix}, \quad (4.6.1)$$

with the bulge located at position $(4, 2)$. The 2×2 subpencil outlined in (4.6.1) is the bulge pencil. Its eigenvalues are ρ and ∞ [69, Chap. 7].[14] If we now do one more transformation on the left, moving the bulge from A to B, we obtain

$$\begin{bmatrix} \times \times \times \times \times \times \\ \times \times \times \times \times \times \\ \times \times \times \times \times \\ \times \times \times \times \\ \times \times \times \\ \times \times \end{bmatrix} \begin{bmatrix} \times \times \times \times \times \times \\ \times \times \times \times \times \\ \times \times \times \times \\ + \times \times \times \\ \times \times \\ \times \end{bmatrix}.$$

This is a Hessenberg pair, and the eigenvalues of the bulge pencil are now in plain sight. In the $(3, 2)$ position we have the pole ∞, and in the $(4, 3)$ position we have a finite pole, which we know to be the shift ρ. What was opaque before is now transparent.

Certain structured problems require algorithms that chase bulges in both directions in order to preserve the structure. The first example of such an algorithm was the Hamiltonian QR algorithm of Byers [18, 19]. Some more recent examples are algorithms for the palindromic and even eigenvalue problems discussed in [42, 48]. Our understanding of the bulge pencil made it possible to explain completely how to pass bulges (and the shifts that they contain) through each other in general in both structured and unstructured cases [68]. It took time and effort to figure this out, but now, in light of what we know about pole swapping, we can see that passing shifts through each other is just a matter of swapping two eigenvalues of the pole pencil. Once again, what was opaque before is now transparent.

---

[14]In [69] we considered (large-bulge) multishift algorithms with k shifts $\rho_1, \ldots, \rho_k$. Then the bulge pencil is $(k+1) \times (k+1)$ and has eigenvalues $\rho_1, \ldots, \rho_k$, and ∞. Here we are considering only the case $k = 1$. We will consider the more general setting in Section 7.3.

4.6. Connections to earlier work

Tightly and optimally packed shifts

The schemes discussed in Section 4.4 insert not just one shift but long chains of shifts $\rho_1, \ldots,$ ρ_m into the pencil as poles and then chase them downward (or upward) in a bunch. In such a scheme it is important for efficiency to have the shifts packed as tightly together as possible. It is clear that in our current scenario we achieve this; the shifts $\rho_1, \ldots, \rho_m$ appear as adjacent poles in the Hessenberg pair, and there is no way they could be packed any closer. (The same result is achieved effortlessly when this methodology is applied to core-chasing algorithms [3].) In contrast, in the bulge-chasing scenario, the packing of bulges is not naturally optimal, and it is not obvious how to fix the problem. However, with some effort a remedy was eventually found [39]. In hindsight we can show that the remedy is a disguised implementation of a pole-swapping algorithm.

We have explained already that pole swapping reduces to bulge chasing if all poles that are not shifts are set to infinity. The philosophy is, however, different. Bulge chasing executes in each step an equivalence where the transforms on left and right act on columns and rows having the same indices, say i and $i+1$. Pole swapping, on the other hand, has the transformation on the left acting on rows i and $i+1$, while the transformation on the right acts on columns $i-1$ and i. Pole swapping is half an equivalence off compared to bulge chasing. This lag is natural in the pole-swapping setting and appears to be the foundational strategy to get optimally packed bulges.

An optimally packed chain of two single shifts in the bulge-chasing setting would, ideally, look like

$$\begin{bmatrix} \times & \times & \times & \times & \times & \times \\ & \times & \times & \times & \times & \times \\ + & \times & \times & \times & \times & \times \\ & + & \times & \times & \times & \times \\ & & & \times & \times & \times \\ & & & & \times & \times \end{bmatrix} \begin{bmatrix} \times & \times & \times & \times & \times & \times \\ & \times & \times & \times & \times & \times \\ & & \times & \times & \times & \times \\ & & & \times & \times & \times \\ & & & & \times & \times \\ & & & & & \times \end{bmatrix}, \qquad (4.6.2)$$

whereas in the pole-swapping setting it would resemble

$$\begin{bmatrix} \times & \times & \times & \times & \times & \times \\ \times & \times & \times & \times & \times & \times \\ & \times & \times & \times & \times & \times \\ & & \times & \times & \times & \times \\ & & & \times & \times & \times \\ & & & & \times & \times \end{bmatrix} \begin{bmatrix} \times & \times & \times & \times & \times & \times \\ + & \times & \times & \times & \times & \times \\ & + & \times & \times & \times & \times \\ & & & \times & \times & \times \\ & & & & \times & \times \\ & & & & & \times \end{bmatrix}.$$

For simplicity, and without loss of generality, we restrict ourselves to two single shifts.

We have seen that getting an optimally packed chain of shifts in the pole-swapping setting is trivial. In the bulge-chasing case, however, it is impossible to achieve (4.6.2). Introducing the first shift and chasing it down a row results in

$$\begin{bmatrix} \times & \times & \times & \times & \times & \times \\ & \times & \times & \times & \times & \times \\ & & \times & \times & \times & \times \\ & + & \times & \times & \times & \times \\ & & & \times & \times & \times \\ & & & & \times & \times \end{bmatrix} \begin{bmatrix} \times & \times & \times & \times & \times & \times \\ & \times & \times & \times & \times & \times \\ & & \times & \times & \times & \times \\ & & & \times & \times & \times \\ & & & & \times & \times \\ & & & & & \times \end{bmatrix}.$$

Introducing the second shift does not work. We end up with

$$\begin{bmatrix} \times & \times & \times & \times & \times & \times \\ & \times & \times & \times & \times & \times \\ + & \times & \times & \times & \times & \times \\ + & + & \times & \times & \times & \times \\ & & & \times & \times & \times \\ & & & & \times & \times \end{bmatrix} \begin{bmatrix} \times & \times & \times & \times & \times & \times \\ & \times & \times & \times & \times & \times \\ & & \times & \times & \times & \times \\ & & & \times & \times & \times \\ & & & & \times & \times \\ & & & & & \times \end{bmatrix},$$

and both single shifts have been combined into a 2×2 multishift bulge. The scheme introduced by Braman, Byers, and Mathias [16] delays the introduction of the second shift until the first has been moved two spots down. This delay is necessary since larger bulges are prone to shift-blurring negatively affecting convergence; see, for instance, [41]. We get

$$\begin{bmatrix} \times & \times & \times & \times & \times & \times \\ \times & \times & \times & \times & \times & \times \\ + & \times & \times & \times & \times & \times \\ & & \times & \times & \times & \times \\ & & + & \times & \times & \times \\ & & & & \times & \times \end{bmatrix} \begin{bmatrix} \times & \times & \times & \times & \times & \times \\ & \times & \times & \times & \times & \times \\ & & \times & \times & \times & \times \\ & & & \times & \times & \times \\ & & & & \times & \times \\ & & & & & \times \end{bmatrix},$$

which are so-called tightly packed shifts. It is impossible to pack them any closer; otherwise the two 2×2 bulge pencils (marked in the figure) would overlap.

A solution to pack the bulges as tightly as in (4.6.2) was proposed by Karlsson, Kressner, and Lang [39]. The trick is to defer some transformations from the right. Suppose the first bulge is introduced and we would like to move it down a row; instead of executing an entire bulge-chasing step, we only execute the transformation from the left—the transformation on the right is postponed. We end up with

$$\begin{bmatrix} \times & \times & \times & \times & \times & \times \\ \times & \times & \times & \times & \times & \times \\ & \times & \times & \times & \times & \times \\ & & \times & \times & \times & \times \\ & & & \times & \times & \times \\ & & & & \times & \times \end{bmatrix} \begin{bmatrix} \times & \times & \times & \times & \times & \times \\ & \times & \times & \times & \times & \times \\ + & \times & \times & \times & \times \\ & & \times & \times & \times \\ & & & \times & \times \\ & & & & \times \end{bmatrix},$$

which is nothing other than having moved the first pole down a position. Next we introduce the second shift, but we do not execute the transformation from the right. We get

$$\begin{bmatrix} \times & \times & \times & \times & \times & \times \\ \times & \times & \times & \times & \times & \times \\ & \times & \times & \times & \times & \times \\ & & \times & \times & \times & \times \\ & & & \times & \times & \times \\ & & & & \times & \times \end{bmatrix} \begin{bmatrix} \times & \times & \times & \times & \times & \times \\ + & \times & \times & \times & \times & \times \\ + & \times & \times & \times & \times \\ & & \times & \times & \times \\ & & & \times & \times \\ & & & & \times \end{bmatrix}.$$

To continue the chasing, we now execute the first of the right transformations that had been delayed. This creates a new bulge in A, which is then annihilated by a transformation from the left. A normal bulge-chasing step would then push the bulge further by a transformation on the right, but again we delay that transformation. We end up with

$$\begin{bmatrix} \times & \times & \times & \times & \times & \times \\ \times & \times & \times & \times & \times & \times \\ & \times & \times & \times & \times & \times \\ & & \times & \times & \times & \times \\ & & & \times & \times & \times \\ & & & & \times & \times \end{bmatrix} \begin{bmatrix} \times & \times & \times & \times & \times & \times \\ + & \times & \times & \times & \times & \times \\ & & \times & \times & \times & \times \\ & & + & \times & \times & \times \\ & & & & \times & \times \\ & & & & & \times \end{bmatrix},$$

after which we can do the same with the second shift. This can now be understood as pole swapping, but the description in terms of bulges and delayed transformations conceals this fact.

Karlsson et al. [39] discussed the optimal packing of the bulges in terms of double-shift bulges. Since we have not discussed double-shift pole-swapping algorithms in this chapter, we do not explore this. The principles are, however, identical. The algorithm of Karlsson et al. [39] is a pole-swapping algorithm (with poles at infinity) avant la lettre.

4.7 • Proof of Theorem 4.3.3

In order to prove the theorem, we will need to develop some machinery.

Preparatory work

Following [25] we introduce two bivariate functions,

$$M(\rho, \sigma) = (A - \rho B)(A - \sigma B)^{-1} \quad \text{and} \quad N(\rho, \sigma) = (A - \sigma B)^{-1}(A - \rho B),$$

where σ is not in the spectrum of (A, B). Remark 4.1.1 remains in effect: an expression of the form $A - \tau B$ stands for an equivalence class of operators $\beta A - \alpha B$, where $\alpha/\beta = \tau$.

If ∞ is not in the spectrum of (A, B), i.e., B is nonsingular, we can consider infinite values of ρ and σ. We have $M(\infty, \sigma) = B(A - \sigma B)^{-1} = (AB^{-1} - \sigma I)^{-1}$ and $M(\rho, \infty) = (A - \rho B)B^{-1} = AB^{-1} - \rho I$. Similarly $N(\infty, \sigma) = (B^{-1}A - \sigma I)^{-1}$ and $N(\rho, \infty) = B^{-1}A - \rho I$.

Simple algebra shows that, for finite ρ and σ,

$$\begin{aligned} M(\rho, \sigma) &= I + (\sigma - \rho)B(A - \sigma B)^{-1}, \\ N(\rho, \sigma) &= I + (\sigma - \rho)(A - \sigma B)^{-1}B. \end{aligned} \quad (4.7.1)$$

The variable ρ appears only in the coefficient $(\sigma - \rho)$, which suggests that ρ might be less important than σ. This turns out to be correct.

The reader can easily check that if B is nonsingular,

$$\begin{aligned} M(\rho, \sigma) &= (AB^{-1} - \rho I)(AB^{-1} - \sigma I)^{-1}, \\ N(\rho, \sigma) &= (B^{-1}A - \sigma I)^{-1}(B^{-1}A - \rho I). \end{aligned} \quad (4.7.2)$$

Moreover, if ρ is not an eigenvalue of (A, B), then

$$M(\rho, \sigma)^{-1} = M(\sigma, \rho) \quad \text{and} \quad N(\rho, \sigma)^{-1} = N(\sigma, \rho).$$

Lemma 4.7.1.

(a) *(Commutativity)*
$$\begin{aligned} M(\rho_1, \sigma_1)M(\rho_2, \sigma_2) &= M(\rho_2, \sigma_2)M(\rho_1, \sigma_1), \\ N(\rho_1, \sigma_1)N(\rho_2, \sigma_2) &= N(\rho_2, \sigma_2)N(\rho_1, \sigma_1). \end{aligned}$$

(b) *(Mobility of ρ)*
$$\begin{aligned} M(\rho_1, \sigma_1)M(\rho_2, \sigma_2) &= M(\rho_2, \sigma_1)M(\rho_1, \sigma_2), \\ N(\rho_1, \sigma_1)N(\rho_2, \sigma_2) &= N(\rho_2, \sigma_1)N(\rho_1, \sigma_2). \end{aligned}$$

Proof. Assume at first that B is nonsingular, and let $C = AB^{-1}$. Then by (4.7.2)

$$M(\rho_1, \sigma_1)M(\rho_2, \sigma_2) = (C - \rho_1 I)(C - \sigma_1 I)^{-1}(C - \rho_2 I)(C - \sigma_2 I)^{-1}.$$

Since all four factors on the right are functions of the same operator C, they all commute with each other. This proves the "M" part of both claims. A similar argument holds for the "N" part with $C = B^{-1}A$.

If B is singular, all equations still hold by a continuity argument. □

Given any *poles* $\sigma_1, \sigma_2, \sigma_3, \ldots$, none in the spectrum of (A, B), and *shifts* $\rho_1, \rho_2, \rho_3, \ldots$, with $\rho_i \neq \sigma_j$ for all i and j, we define a nested sequence of *rational Krylov subspaces*

$$\mathcal{K}_1(A, B, v, [\,]) = \operatorname{span}\{v\},$$
$$\mathcal{K}_2(A, B, v, [\sigma_1]) = \operatorname{span}\{v, M(\rho_1, \sigma_1)v\},$$
$$\mathcal{K}_3(A, B, v, [\sigma_1, \sigma_2]) = \operatorname{span}\{v, M(\rho_1, \sigma_1)v, M(\rho_2, \sigma_2)M(\rho_1, \sigma_1)v\},$$

and in general

$$\mathcal{K}_j(A, B, v, [\sigma_1, \ldots, \sigma_{j-1}]) = \operatorname{span}\left\{v, M(\rho_1, \sigma_1)v, \ldots, \left(\prod_{i=1}^{j-1} M(\rho_i, \sigma_i)\right)v\right\}.$$

The product $\prod_{i=1}^{j-1} M(\rho_i, \sigma_i)$ is well defined because the factors commute. We will see in a minute that this definition is independent of the choice of the shifts $\rho_1, \rho_2, \ldots$. A simpler development that does not make use of these shifts is laid out in [21], but it assumes throughout that B is nonsingular.

The poles of $\mathcal{K}_j$ in this definition are generally not the same as the poles of the pair (A, B).

We are using the symbol $\mathcal{K}_j$ to denote both standard and rational Krylov subspaces. The meaning in each case is easily deduced from the number and type of arguments.

We define another nested sequence of *rational Krylov subspaces* by

$$\mathcal{L}_j(A, B, v, [\sigma_1, \ldots, \sigma_{j-1}]) = \operatorname{span}\left\{v, N(\rho_1, \sigma_1)v, \ldots, \left(\prod_{i=1}^{j-1} N(\rho_i, \sigma_i)\right)v\right\}.$$

We will see that this definition is also independent of $\rho_1, \rho_2, \ldots$.

In the following results we will make the blanket assumption that none of $\rho_1, \rho_2, \rho_3, \ldots$ are poles of the Hessenberg pair (A, B), nor are they eigenvalues. For simplicity we will not repeat this assumption in the statement of each result.

Lemma 4.7.2. *For any $w \in \mathbb{C}^n$ and any shifts ρ and $\hat{\rho}$ distinct from σ,*

$$\operatorname{span}\{w, M(\rho, \sigma)w\} = \operatorname{span}\{w, M(\hat{\rho}, \sigma)w\},$$

$$\operatorname{span}\{w, N(\rho, \sigma)w\} = \operatorname{span}\{w, N(\hat{\rho}, \sigma)w\}.$$

Proof. If ρ, $\hat{\rho}$, and σ are all finite, $M(\rho, \sigma)w = w + (\sigma - \rho)B(A - \sigma B)^{-1}w$ by (4.7.1), so

$$\operatorname{span}\{w, M(\rho, \sigma)w\} = \operatorname{span}\{w, B(A - \sigma B)^{-1}w\} = \operatorname{span}\{w, M(\hat{\rho}, \sigma)w\}.$$

This equation also holds for $\rho = \infty$ or $\hat{\rho} = \infty$ because $M(\infty, \sigma) = B(A - \sigma B)^{-1}$. For $\sigma = \infty$, $M(\rho, \infty)w = (A - \rho B)B^{-1}w = AB^{-1}w - \rho w$, so

$$\operatorname{span}\{w, M(\rho, \infty)w\} = \operatorname{span}\{w, AB^{-1}w\} = \operatorname{span}\{w, M(\hat{\rho}, \infty)\}.$$

This proves the "M" result. The "N" result is proved similarly. $\square$

We now justify our claim that rational Krylov subspaces do not depend on the choice of shifts $\rho_1, \rho_2, \rho_3, \ldots$.

4.7. Proof of Theorem 4.3.3

Theorem 4.7.3. *For any $\hat{\rho}$ distinct from σ_1, σ_2, σ_3, ..., and for $j = 1, 2, 3, \ldots$*

$$\mathrm{span}\left\{v, M(\rho_1,\sigma_1)v, M(\rho_2,\sigma_2)M(\rho_1,\sigma_1)v, \ldots, \left(\prod_{i=1}^{j-1} M(\rho_i,\sigma_i)\right)v\right\}$$
$$= \mathrm{span}\left\{v, M(\hat{\rho},\sigma_1)v, M(\hat{\rho},\sigma_2)M(\hat{\rho},\sigma_1)v, \ldots, \left(\prod_{i=1}^{j-1} M(\hat{\rho},\sigma_i)\right)v\right\}$$

and

$$\mathrm{span}\left\{v, N(\rho_1,\sigma_1)v, N(\rho_2,\sigma_2)N(\rho_1,\sigma_1)v, \ldots, \left(\prod_{i=1}^{j-1} N(\rho_i,\sigma_i)\right)v\right\}$$
$$= \mathrm{span}\left\{v, N(\hat{\rho},\sigma_1)v, N(\hat{\rho},\sigma_2)N(\hat{\rho},\sigma_1)v, \ldots, \left(\prod_{i=1}^{j-1} N(\hat{\rho},\sigma_i)\right)v\right\}.$$

In words, we can replace any ρ_1, ρ_2, ρ_3, ... by a single scalar $\hat{\rho}$. This shows that the definitions of the rational Krylov subspaces are independent of the ρ_i.

Proof. We'll show how to prove the "M" part and leave the "N" part as an exercise for the reader. The proof is by induction on j. The case $j = 1$ is trivial. Case $j = 2$ is Lemma 4.7.2. Case $j = 3$ tells the whole story. Temporarily define

$$\mathcal{U}_2 = \mathrm{span}\{v, M(\rho_1,\sigma_1)v\}, \qquad \widehat{\mathcal{U}}_2 = \mathrm{span}\{v, M(\hat{\rho},\sigma_1)v\},$$

$$\mathcal{U}_3 = \mathrm{span}\{v, M(\rho_1,\sigma_1)v, M(\rho_2,\sigma_2)M(\rho_1,\sigma_1)v\},$$

and

$$\widehat{\mathcal{U}}_3 = \mathrm{span}\{v, M(\hat{\rho},\sigma_1)v, M(\hat{\rho},\sigma_2)M(\hat{\rho},\sigma_1)v\}.$$

We know from the case $j = 2$ that $\mathcal{U}_2 = \widehat{\mathcal{U}}_2$, and we have to show that $\mathcal{U}_3 = \widehat{\mathcal{U}}_3$. First of all

$$\mathcal{U}_3 = \mathcal{U}_2 + \mathrm{span}\{M(\rho_2,\sigma_2)M(\rho_1,\sigma_1)v\}.$$

We know that $\mathcal{U}_2 = \widehat{\mathcal{U}}_2$, so we just need to show that we can replace the additional term by $\mathrm{span}\{M(\hat{\rho},\sigma_2)M(\hat{\rho},\sigma_1)v\}$. We will use Lemma 4.7.2 twice, first to replace ρ_2, then ρ_1, by $\hat{\rho}$. Let $w = M(\rho_1,\sigma_1)v$. Then $w \in \widehat{\mathcal{U}}_2$, so

$$\mathcal{U}_3 = \widehat{\mathcal{U}}_2 + \mathrm{span}\{M(\rho_2,\sigma_2)w\} = \widehat{\mathcal{U}}_2 + \mathrm{span}\{w, M(\rho_2,\sigma_2)w\}.$$

Applying Lemma 4.7.2 to the last term, we get

$$\mathcal{U}_3 = \widehat{\mathcal{U}}_2 + \mathrm{span}\{w, M(\hat{\rho},\sigma_2)w\} = \widehat{\mathcal{U}}_2 + \mathrm{span}\{M(\hat{\rho},\sigma_2)w\}.$$

$M(\hat{\rho},\sigma_2)w = M(\hat{\rho},\sigma_2)M(\rho_1,\sigma_1)v = M(\rho_1,\sigma_2)M(\hat{\rho},\sigma_1)v$ by the definition of w and part (b) of Lemma 4.7.1, which allows us to swap $\hat{\rho}$ and ρ_1. Now let $x = M(\hat{\rho},\sigma_1)v$. Since $x \in \widehat{\mathcal{U}}_2$,

$$\mathcal{U}_3 = \widehat{\mathcal{U}}_2 + \mathrm{span}\{M(\rho_1,\sigma_2)x\} = \widehat{\mathcal{U}}_2 + \mathrm{span}\{x, M(\rho_1,\sigma_2)x\}.$$

Applying Lemma 4.7.2 a second time, we get

$$\mathcal{U}_3 = \widehat{\mathcal{U}}_2 + \mathrm{span}\{x, M(\hat{\rho},\sigma_2)x\} = \widehat{\mathcal{U}}_2 + \mathrm{span}\{M(\hat{\rho},\sigma_2)M(\hat{\rho},\sigma_1)v\} = \widehat{\mathcal{U}}_3,$$

which is the desired result.

Case $j = 4$ is just like $j = 3$, except that Lemma 4.7.2 is used three times, and so on. We leave the general induction step as an exercise for the reader. □

Theorem 4.7.4.

(a) *For any two shifts $\rho \neq \hat{\rho}$, and for $j = 1, 2, 3, \ldots$*

$$\mathcal{K}_j(A, B, v, [\sigma_1, \ldots, \sigma_{j-1}]) = \left(\prod_{i=1}^{j-1} M(\rho, \sigma_i)\right) \mathcal{K}_j(M(\hat{\rho}, \rho), v)$$
$$= \mathcal{K}_j(M(\hat{\rho}, \rho), w),$$
$$\text{where} \quad w = \left(\prod_{i=1}^{j-1} M(\rho, \sigma_i)\right) v;$$

$$\mathcal{L}_j(A, B, v, [\sigma_1, \ldots, \sigma_{j-1}]) = \left(\prod_{i=1}^{j-1} N(\rho, \sigma_i)\right) \mathcal{K}_j(N(\hat{\rho}, \rho), v)$$
$$= \mathcal{K}_j(N(\hat{\rho}, \rho), x),$$
$$\text{where} \quad x = \left(\prod_{i=1}^{j-1} N(\rho, \sigma_i)\right) v.$$

This shows that every rational Krylov subspace is an ordinary Krylov subspace associated with a different starting vector.

(b) *For any shift ρ, for $j = 1, 2, 3, \ldots$*

$$M(\rho, \sigma_j) \mathcal{K}_j(A, B, v, [\sigma_1, \ldots, \sigma_{j-1}]) \subseteq \mathcal{K}_{j+1}(A, B, v, [\sigma_1, \ldots, \sigma_j]),$$
$$N(\rho, \sigma_j) \mathcal{L}_j(A, B, v, [\sigma_1, \ldots, \sigma_{j-1}]) \subseteq \mathcal{L}_{j+1}(A, B, v, [\sigma_1, \ldots, \sigma_j]).$$

Proof. We'll prove the $\mathcal{K}_j$ claims, leaving the $\mathcal{L}_j$ claims for the reader.

(a) We start by recalling the definition of $\mathcal{K}_j$. We then factor out the product $\prod_{i=1}^{j-1} M(\rho, \sigma_i)$, making use of the inverse property $M(\rho, \sigma_i) M(\sigma_i, \rho) = I$ and the commutativity of the factors:

$$\mathcal{K}_j = \mathcal{K}_j(A, B, v, [\sigma_1, \ldots, \sigma_{j-1}])$$
$$= \text{span}\left\{v, M(\rho, \sigma_1)v, \ldots, \left(\prod_{i=1}^{j-1} M(\rho, \sigma_i)\right) v\right\}$$
$$= \left(\prod_{i=1}^{j-1} M(\rho, \sigma_i)\right) \text{span}\left\{\left(\prod_{i=1}^{j-1} M(\sigma_i, \rho)\right) v, \ldots, M(\sigma_{j-1}, \rho)v, v\right\}.$$

Thus

$$\mathcal{K}_j = \mathcal{K}_j(A, B, v, [\sigma_1, \ldots, \sigma_{j-1}])$$
$$= \left(\prod_{i=1}^{j-1} M(\rho, \sigma_i)\right) \text{span}\left\{v, M(\sigma_{j-1}, \rho)v, \ldots, \left(\prod_{i=1}^{j-1} M(\sigma_i, \rho)\right) v\right\}$$
$$= \left(\prod_{i=1}^{j-1} M(\rho, \sigma_i)\right) \text{span}\left\{v, M(\hat{\rho}, \rho)v, \ldots, \left(\prod_{i=1}^{j-1} M(\hat{\rho}, \rho)\right) v\right\}.$$

Here we have used Theorem 4.7.3 to replace $\sigma_{j-1}, \ldots, \sigma_1$ by $\hat{\rho}$. This last space is an ordinary Krylov subspace

$$\left(\prod_{i=1}^{j-1} M(\rho, \sigma_i)\right) \mathcal{K}_j(M(\hat{\rho}, \rho), v) = \mathcal{K}_j(M(\hat{\rho}, \rho), w).$$

This proves part (a).

(b) From part (a) we have

$$\mathcal{K}_j(A, B, v, [\sigma_1, \ldots, \sigma_{j-1}]) = \mathcal{K}_j(M(\hat{\rho}, \rho), w)$$

and

$$\mathcal{K}_{j+1}(A, B, v, [\sigma_1, \ldots, \sigma_j]) = \mathcal{K}_{j+1}(M(\hat{\rho}, \rho), \tilde{w}),$$

4.7. Proof of Theorem 4.3.3

where $\tilde{w} = \left(\prod_{i=1}^{j} M(\rho, \sigma_i)\right) v = M(\rho, \sigma_j) w$. Therefore

$$M(\rho, \sigma_j) \mathcal{K}_j = M(\rho, \sigma_j) \mathcal{K}_j(M(\hat{\rho}, \rho), w) = \mathcal{K}_j(M(\hat{\rho}, \rho), \tilde{w}) \subseteq \mathcal{K}_{j+1}(M(\hat{\rho}, \rho), \tilde{w}).$$

This proves part (b). □

Lemma 4.7.5. *Let (A, B) be a proper Hessenberg pair, $A = [a_1, \ldots, a_n]$, $B = [b_1, \ldots, b_n]$. Then for $j = 1, \ldots, n-1$*

$$\mathrm{span}\{a_1, \ldots, a_j\} \neq \mathrm{span}\{b_1, \ldots, b_j\}.$$

Proof. As part of the definition of proper Hessenberg pair, we have $\mathrm{span}\{a_1\} \neq \mathrm{span}\{b_1\}$. This takes care of the case $j = 1$. Now make the induction hypothesis that for some $j < n$, $\mathrm{span}\{a_1, \ldots, a_{j-1}\} \neq \mathrm{span}\{b_1, \ldots, b_{j-1}\}$, and we will prove by contradiction that $\mathrm{span}\{a_1, \ldots, a_j\} \neq \mathrm{span}\{b_1, \ldots, b_j\}$. Assume therefore that $\mathrm{span}\{a_1, \ldots, a_j\} = \mathrm{span}\{b_1, \ldots, b_j\}$. Then there is a $j \times j$ matrix C such that

$$[a_1, \ldots, a_j] = [b_1, \ldots, b_j] \begin{bmatrix} c_{11} & \cdots & c_{1j} \\ \vdots & & \vdots \\ c_{j1} & \cdots & c_{jj} \end{bmatrix}.$$

By the induction hypothesis there is a column a_i with $i \leq j-1$ such that $a_i \notin \mathrm{span}\{b_1, \ldots, b_{j-1}\}$.[15] This implies $c_{ji} \neq 0$. By the Hessenberg form of both A and B,

$$0 = a_{j+1,i} = \sum_{k=1}^{j} b_{j+1,k} c_{ki} = b_{j+1,j} c_{ji}.$$

Thus $b_{j+1,j} = 0$, which also forces $a_{j+1,j} = 0$, since

$$a_{j+1,j} = \sum_{k=1}^{j} b_{j+1,k} c_{kj} = b_{j+1,j} c_{jj} = 0.$$

This contradicts the proper upper Hessenberg form of (A, B). □

Proposition 4.7.6. *Let (A, B) be a proper upper Hessenberg pair with ordered pole set $[\sigma_1, \ldots, \sigma_{n-1}]$. Let $\mathcal{E}_j = \mathrm{span}\{e_1, \ldots, e_j\}$, as before. Then*

$$\mathcal{E}_j = \mathcal{K}_j(A, B, e_1, [\sigma_1, \ldots, \sigma_{j-1}]), \quad j = 1, \ldots, n-1,$$

$$\mathcal{E}_j = \mathcal{L}_j(A, B, e_1, [\sigma_2, \ldots, \sigma_j]), \quad j = 1, \ldots, n-2.$$

Notice that in the $\mathcal{L}_j$ spaces the poles are $[\sigma_2, \ldots, \sigma_j]$, starting from σ_2.

Proof. First we prove the claim about $\mathcal{K}_j$. For $j = 1$ the equality is trivially true. Now we assume for proof by induction that $\mathcal{E}_j = \mathcal{K}_j$, and we will show that this implies that $\mathcal{E}_{j+1} = \mathcal{K}_{j+1}$. Clearly $\mathcal{E}_j = \mathcal{K}_j \subseteq \mathcal{K}_{j+1}$, so we just need to show that $e_{j+1} \in \mathcal{K}_{j+1}$. For this it suffices to show that $\mathcal{K}_{j+1}$ contains a $w \in \mathcal{E}_{j+1}$ such that $w_{j+1} \neq 0$. From part (b) of Theorem 4.7.4 we know that $M(\rho, \sigma_j) \mathcal{E}_j \subseteq \mathcal{K}_{j+1}$ for any ρ. Thus, for every $x \in \mathcal{E}_j$,

$$(A - \rho B)(A - \sigma_j B)^{-1} x = M(\rho, \sigma_j) x \in \mathcal{K}_{j+1}.$$

Now let $x = (A - \sigma_j B) e_j$. Because σ_j is the jth pole of (A, B), $a_{j+1,j} - \sigma_j b_{j+1,j} = 0$. Therefore $A - \sigma_j B$ is block triangular with a $j \times j$ leading block, from which it follows that $x \in \mathcal{E}_j$.

---
[15] To achieve this, we reverse the roles of A and B if necessary.

Now let $w = (A - \rho B)e_j$. Since $A - \rho B$ is upper Hessenberg, $w \in \mathcal{E}_{j+1}$. Since ρ is (by our blanket assumption) not a pole of (A, B), $A - \rho B$ is *properly* upper Hessenberg, so $w_{j+1} \neq 0$. Finally
$$w = (A - \rho B)e_j = (A - \rho B)(A - \sigma_j B)^{-1}x = M(\rho, \sigma_j)x.$$
We have found a $w \in \mathcal{K}_{j+1}$ such that $w \in \mathcal{E}_{j+1}$ and $w_{j+1} \neq 0$. This proves the $\mathcal{K}_j$ claim.

Now we move on to the $\mathcal{L}_j$ claim. The proof is similar but a bit trickier. Assume for proof by induction that $\mathcal{E}_j = \mathcal{L}_j = \mathcal{L}_j(A, B, e_1, [\sigma_2, \ldots, \sigma_j])$; we will show that this implies $\mathcal{E}_{j+1} = \mathcal{L}_{j+1} = \mathcal{L}_{j+1}(A, B, e_1, [\sigma_2, \ldots, \sigma_{j+1}])$. Don't forget that the poles start from σ_2 in this case. Clearly $\mathcal{E}_j = \mathcal{L}_j \subseteq \mathcal{L}_{j+1}$, so we just need to show that $\mathcal{L}_{j+1}$ contains a vector $w \in \mathcal{E}_{j+1}$ such that $w_{j+1} \neq 0$. By part (b) of Theorem 4.7.4, $N(\rho, \sigma_{j+1})\mathcal{E}_j \subseteq \mathcal{L}_{j+1}$, so for every $x \in \mathcal{E}_j$,
$$(A - \sigma_{j+1}B)^{-1}(A - \rho B)x = N(\rho, \sigma_{j+1})x \in \mathcal{L}_{j+1}.$$
To complete the proof, we just need to find x and w such that
$$w = (A - \sigma_{j+1}B)^{-1}(A - \rho B)x, \quad x \in \mathcal{E}_j, \quad w \in \mathcal{E}_{j+1} \setminus \mathcal{E}_j.$$
For any $x \in \mathcal{E}_j$, let $y = (A - \rho B)x$, so that $w = (A - \sigma_{j+1}B)^{-1}y$. Since $A - \rho B$ is upper Hessenberg, $y \in \mathcal{E}_{j+1}$. Since σ_{j+1} is a pole of (A, B), $a_{j+2,j+1} - \sigma_{j+1}b_{j+2,j+1} = 0$. Thus $A - \sigma_{j+1}B$ is block triangular with a leading $(j+1) \times (j+1)$ block, and the same is true of its inverse. Therefore $w = (A - \sigma_{j+1}B)^{-1}y \in \mathcal{E}_{j+1}$. Now we just have to show that there is an x that delivers a w for which $w_{j+1} \neq 0$.

Let us assume that this is not true, which would mean that for every $x \in \mathcal{E}_j$, the corresponding w also lies in $\mathcal{E}_j$. Since $A - \rho B$ is nonsingular by assumption, this would imply that
$$(A - \sigma_{j+1}B)^{-1}(A - \rho B)\mathcal{E}_j = \mathcal{E}_j.$$
We will show that this leads to a contradiction. Make the abbreviations $C = A - \sigma_{j+1}B$ and $D = A - \rho B$. Then our assumption is equivalent to
$$D\mathcal{E}_j = C\mathcal{E}_j. \tag{4.7.3}$$
Since ρ is (by assumption) not a pole of (A, B), $A - \rho B$ is properly upper Hessenberg. Therefore (C, D) is a proper Hessenberg pair. Letting
$$\mathcal{C}_j = C\mathcal{E}_j = \operatorname{span}\{c_1, \ldots, c_j\} \quad \text{and} \quad \mathcal{D}_j = D\mathcal{E}_j = \operatorname{span}\{d_1, \ldots, d_j\},$$
we note that (4.7.3) implies that $\operatorname{span}\{c_1, \ldots, c_j\} = \operatorname{span}\{d_1, \ldots, d_j\}$. On the other hand, Lemma 4.7.5 implies that $\operatorname{span}\{c_1, \ldots, c_j\} \neq \operatorname{span}\{d_1, \ldots, d_j\}$. This is a contradiction. □

Proof of the theorem

With this machinery in hand, we can now prove Theorem 4.3.3. Let us recall the situation. Each move of type I or II is an equivalence transform of the form
$$\hat{A} = Q_j^* A Z_{j-1}, \quad \hat{B} = Q_j^* B Z_{j-1}.$$
The case $j = 1$ denotes a move of type I, and we have $Z_0 = I$. The case $j = n$ also denotes a type I move, and in this case $Q_n = I$. The cases $j = 2, \ldots, n-1$ are of type II. Suppose (A, B) has poles $\sigma_1, \ldots, \sigma_{n-1}$. A move of type II interchanges poles σ_{j-1} and σ_j. For the moves of type I, in the case $j = 1$, suppose the pole σ_1 is replaced by a new pole σ_0; in the case $j = n$, suppose σ_{n-1} is replaced by a new pole σ_n.

4.7. Proof of Theorem 4.3.3

We define sequences of nested subspaces $(\mathcal{Q}_k)$ and $(\mathcal{Z}_k)$, where $\mathcal{Q}_k$ (resp., $\mathcal{Z}_k$) is the space spanned by the first k columns of Q_j (resp., Z_{j-1}). Because Q_j and Z_{j-1} are core transformations, these spaces are mostly trivial in this setting: $\mathcal{Q}_k = \mathcal{E}_k$ except when $k = j$, and $\mathcal{Z}_k = \mathcal{E}_k$ except when $k = j - 1$.

We have to show that the transformation
$$\hat{A} - \lambda \hat{B} = Q_j^*(A - \lambda B)Z_{j-1}$$
effects the subspace iterations
$$\mathcal{Q}_j = (A - \sigma_{j-1}B)(A - \sigma_j B)^{-1}\mathcal{E}_j$$
and
$$\mathcal{Z}_{j-1} = (A - \sigma_j B)^{-1}(A - \sigma_{j-1}B)\mathcal{E}_{j-1}.$$

Proof. Proposition 4.1.2 shows that $Q_1 e_1 = \delta (A - \sigma_0 B)(A - \sigma_1 B)^{-1} e_1$ for some nonzero δ. This establishes the case $j = 1$ of Theorem 4.3.3.

Now consider $j > 1$. The transformation $\hat{A} - \lambda \hat{B} = Q_j^*(A - \lambda B)Z_{j-1}$ interchanges poles σ_{j-1} and σ_j, so the ordered pole set of $(\hat{A}, \hat{B})$ is
$$[\sigma_1, \ldots, \sigma_{j-2}, \sigma_j, \sigma_{j-1}, \sigma_{j+1}, \ldots, \sigma_{n-1}].$$
Applying Proposition 4.7.6 to $(\hat{A}, \hat{B})$ we have
$$\mathcal{E}_j = \mathcal{K}_j(\hat{A}, \hat{B}, e_1, [\sigma_1, \ldots, \sigma_{j-2}, \sigma_j]).$$
Therefore
$$\mathcal{Q}_j = Q_j \mathcal{E}_j = Q_j \mathcal{K}_j(\hat{A}, \hat{B}, e_1, [\sigma_1, \ldots, \sigma_{j-2}, \sigma_j])$$
$$= \mathcal{K}_j(A, B, Q_j e_1, [\sigma_1, \ldots, \sigma_{j-2}, \sigma_j]).$$
Noting that $Q_j e_1 = e_1$, and using part (a) of Theorem 4.7.4 twice, we obtain
$$\mathcal{Q}_j = \mathcal{K}_j(A, B, e_1, [\sigma_1, \ldots, \sigma_{j-2}, \sigma_j])$$
$$= M(\rho, \sigma_j) \left(\prod_{i=1}^{j-2} M(\rho, \sigma_i)\right) \mathcal{K}_j(M(\hat{\rho}, \rho), e_1)$$
$$= M(\sigma_{j-1}, \rho) M(\rho, \sigma_j) \left(\prod_{i=1}^{j-1} M(\rho, \sigma_i)\right) \mathcal{K}_j(M(\hat{\rho}, \rho), e_1)$$
$$= M(\sigma_{j-1}, \sigma_j) \mathcal{K}_j(A, B, e_1, [\sigma_1, \ldots, \sigma_{j-2}, \sigma_{j-1}])$$
$$= (A - \sigma_{j-1}B)(A - \sigma_j B)^{-1}\mathcal{E}_j.$$

In the final step we used Proposition 4.7.6 again. We have obtained the desired result: $\mathcal{Q}_j = (A - \sigma_{j-1}B)(A - \sigma_j B)^{-1}\mathcal{E}_j$.

Now consider the spaces $\mathcal{Z}_{j-1}$. In the case $j = 2$ we have
$$\hat{A} - \lambda \hat{B} = Q_2^*(A - \lambda B)Z_1.$$
Substituting $\lambda = \sigma_2$ and solving for Z_1, we have
$$Z_1 = (A - \sigma_2 B)^{-1} Q_2 (\hat{A} - \sigma_2 \hat{B}).$$
The ordered pole set for $(\hat{A}, \hat{B})$ is $[\sigma_2, \sigma_1, \sigma_3, \ldots, \sigma_{n-1}]$, so $(\hat{A} - \sigma_2 \hat{B}) e_1 = \gamma e_1$ for some nonzero γ. Similarly $(A - \sigma_1 B) e_1 = \delta e_1$ for some nonzero δ. Therefore
$$Z_1 e_1 = \gamma (A - \sigma_2 B)^{-1} Q_2 e_1 = \gamma (A - \sigma_2 B)^{-1} e_1 = \gamma \delta^{-1} (A - \sigma_2 B)^{-1}(A - \sigma_1 B) e_1.$$

This proves that
$$\mathcal{Z}_1 = (A - \sigma_2 B)^{-1}(A - \sigma_1 B)\mathcal{E}_1,$$
as desired.

For $j > 2$ we have $\hat{A} - \lambda\hat{B} = Q_j^*(A - \lambda B)Z_{j-1}$. Arguing just as we did for $\mathcal{Q}_j$, we have
$$\mathcal{Z}_{j-1} = Z_{j-1}\mathcal{E}_{j-1} = Z_{j-1}\mathcal{L}_{j-1}(\hat{A}, \hat{B}, e_1, [\sigma_2, \ldots, \sigma_{j-2}, \sigma_j])$$
$$= \mathcal{L}_{j-1}(A, B, Z_{j-1}e_1, [\sigma_2, \ldots, \sigma_{j-2}, \sigma_j]).$$

Using $Z_{j-1}e_1 = e_1$, and invoking part (a) of Theorem 4.7.4 twice, we obtain
$$\mathcal{Z}_{j-1} = \mathcal{L}_{j-1}(A, B, e_1, [\sigma_2, \ldots, \sigma_{j-2}, \sigma_j])$$
$$= N(\rho, \sigma_j)\left(\prod_{i=2}^{j-2} N(\rho, \sigma_i)\right)\mathcal{K}_{j-1}(N(\hat{\rho}, \rho), e_1)$$
$$= N(\sigma_{j-1}, \rho)N(\rho, \sigma_j)\left(\prod_{i=2}^{j-1} N(\rho, \sigma_i)\right)\mathcal{K}_{j-1}(N(\hat{\rho}, \rho), e_1)$$
$$= N(\sigma_{j-1}, \sigma_j)\mathcal{L}_{j-1}(A, B, e_1, [\sigma_2, \ldots, \sigma_j])$$
$$= (A - \sigma_j B)^{-1}(A - \sigma_{j-1}B)\mathcal{E}_{j-1}.$$

This is the desired result. □

Remark 4.7.7. *We used Proposition 4.1.2 to prove the case $j = 1$, but we did not use Proposition 4.1.4. In connection with this we remark that Theorem 4.3.3 immediately implies the dual results*
$$\mathcal{Q}_j^\perp = (A^* - \overline{\sigma}_{j-1}B^*)^{-1}(A^* - \overline{\sigma}_j B^*)\mathcal{E}_j^\perp$$
and
$$\mathcal{Z}_{j-1}^\perp = (A^* - \overline{\sigma}_j B^*)(A^* - \overline{\sigma}_{j-1}B^*)^{-1}\mathcal{E}_{j-1}^\perp,$$
obtained by noting that, for any nonsingular matrix C, $\mathcal{U} = C\mathcal{S}$ if and only if $\mathcal{U}^\perp = (C^*)^{-1}\mathcal{S}^\perp$. We could equally well have derived the dual results first and then deduced Theorem 4.3.3. In that case we would use Proposition 4.1.4 to prove the case $j = n$, and not use Proposition 4.1.2 at all. From Proposition 4.1.4 with $\tau = \sigma_n$ we have immediately
$$Z_{n-1}e_n = \overline{\delta}\,(A^* - \overline{\sigma}_n B^*)(A^* - \overline{\sigma}_{n-1}B^*)^{-1}e_n,$$
which implies
$$\mathcal{Z}_{n-1}^\perp = (A^* - \overline{\sigma}_n B^*)(A^* - \overline{\sigma}_{n-1}B^*)^{-1}\mathcal{E}_{n-1}^\perp,$$
the case $j = n$ of the dual result.

Exercises 4.7

1. Prove Theorem 4.7.3 by induction on j.

Chapter 5
The Standard Eigenvalue Problem

Up to this point we have focused on the generalized eigenvalue problem for a matrix pencil $A - \lambda B$. It is natural to ask whether pole-swapping methods can be used to solve the standard eigenvalue problem for a single matrix $A \in \mathbb{C}^{n \times n}$. This is a special case of the generalized eigenvalue problem, where the pencil is $A - \lambda I$. Let us assume that A has already been reduced to Hessenberg form, so the pair (A, I) is a Hessenberg pair. If we now apply the basic algorithm from Chapter 4 to this pair, the identity matrix will not be preserved, as the transformations are unitary equivalences but not similarities. However the transformed matrix will have some useful properties. For one, it will remain upper Hessenberg. Moreover it will be unitary because I is unitary and so are all of the transformations. Thus we will have to deal with Hessenberg pairs (A, U) for which U is unitary. This may look like a losing strategy, since it forces us to work with two matrices instead of one. It turns out not to be a loser because unitary Hessenberg matrices can be stored in a compact factored form requiring only $O(n)$ storage and can be manipulated almost for free.

5.1 ▪ Unitary Hessenberg matrices

We pause to recall the compact storage scheme for unitary Hessenberg matrices. Let U be a unitary upper Hessenberg matrix:

$$U = \begin{bmatrix} \times & \times & \times & \times & \times & \times \\ \times & \times & \times & \times & \times & \times \\ & \times & \times & \times & \times & \times \\ & & \times & \times & \times & \times \\ & & & \times & \times & \times \\ & & & & \times & \times \end{bmatrix}.$$

We will show that U can be stored as a product of $n-1$ core transformations. We will construct our cores as specified in Section 1.4. Thus each core will have determinant 1 and will produce a vector $\begin{bmatrix} r \\ 0 \end{bmatrix}$ with r real and nonnegative. We begin by applying a core transformation U_1^* to rows 1 and 2 to transform u_{21} to zero:

$$\overset{\curvearrowright}{\begin{bmatrix} \times & \times & \times & \times & \times & \times \\ \times & \times & \times & \times & \times & \times \\ & \times & \times & \times & \times & \times \\ & & \times & \times & \times & \times \\ & & & \times & \times & \times \\ & & & & \times & \times \end{bmatrix}} = \begin{bmatrix} \times & \times & \times & \times & \times & \times \\ & \times & \times & \times & \times & \times \\ & \times & \times & \times & \times & \times \\ & & \times & \times & \times & \times \\ & & & \times & \times & \times \\ & & & & \times & \times \end{bmatrix}.$$

Since the resulting matrix is unitary, its first column must have norm 1. Therefore the $(1,1)$ entry must be 1, provided the core is constructed as specified in Section 1.4. The first row must also have norm 1, implying that all of the other numbers in the first row must be zero:

$$U_1^* U = \begin{bmatrix} \times\times\times\times\times\times \\ \times\times\times\times\times\times \\ \times\times\times\times\times \\ \times\times\times\times \\ \times\times\times \\ \times\times \end{bmatrix} = \begin{bmatrix} 1 \\ & \times\times\times\times \\ & \times\times\times\times \\ & \times\times\times\times \\ & \times\times\times \\ & \times\times \end{bmatrix}.$$

Next we apply a core transformation U_2^* to $U_1^* U$ to annihilate u_{32}. Reasoning as in the first step, we find that

$$U_2^* U_1^* U = \begin{bmatrix} \times\times\times\times\times\times \\ \times\times\times\times\times\times \\ \times\times\times\times\times \\ \times\times\times\times \\ \times\times\times \\ \times\times \end{bmatrix} = \begin{bmatrix} 1 \\ & 1 \\ & & \times\times\times\times \\ & & \times\times\times\times \\ & & \times\times\times \\ & & \times\times \end{bmatrix}.$$

Continuing the process we can transform U to diagonal form:

$$\begin{bmatrix} \times\times\times\times\times\times \\ \times\times\times\times\times\times \\ \times\times\times\times\times \\ \times\times\times\times \\ \times\times\times \\ \times\times \end{bmatrix} = \begin{bmatrix} 1 \\ & 1 \\ & & 1 \\ & & & 1 \\ & & & & 1 \\ & & & & & \times \end{bmatrix}.$$

The last entry in the diagonal matrix is not necessarily 1, but its absolute value clearly is. You can easily show that that entry is the determinant of U. In the general $n \times n$ case we have

$$U_{n-1}^* U_{n-2}^* \cdots U_2^* U_1^* U = D,$$

where $D = \text{diag}\{1, \ldots, 1, d_n\}$ with $d_n = \det(U)$ and $|d_n| = 1$. In our application we are going start with $U = I$, which has determinant 1. All of the cores which will be applied subsequently also have determinant 1, so the condition $\det(U) = 1$ is preserved. Therefore $d_n = 1$, and we can forget about D:

$$U_{n-1}^* U_{n-2}^* \cdots U_2^* U_1^* U = I.$$

If we now invert the core transformations, we get

$$U = U_1 U_2 \cdots U_{n-1}. \tag{5.1.1}$$

This is an extremely compact way of representing U, as each core is determined by just two numbers. Therefore the storage requirement is only $O(n)$ instead of the $O(n^2)$ that is usually required for an $n \times n$ matrix. Pictorially we have

$$U = \begin{bmatrix} \times\times\times\times\times\times \\ \times\times\times\times\times\times \\ \times\times\times\times\times \\ \times\times\times\times \\ \times\times\times \\ \times\times \end{bmatrix} = \text{(product of cores)}.$$

5.2. The RQR algorithm

This efficient way of storing unitary upper Hessenberg matrices has been used by Gragg and others to compute the eigenvalues of unitary upper Hessenberg matrices in a series of papers starting with [32], and more recently by Aurentz et al. [7]. Contrary to our preferred choice of rotations, Gragg used reflections as core transformations and called the resulting $2n$ parameters Schur parameters [32].

Exercises 5.1

1. Show that in the factorization (5.1.1), $\det(U) = d_n$, provided that the core transformations are constructed as described in Section 1.4.

5.2 ▪ The RQR algorithm

We are now equipped to discuss the application of the basic algorithm from Chapter 4 to a Hessenberg pair (A, U), where U is unitary with $\det(U) = 1$, stored in the form (5.1.1). We call this the RQR algorithm, and we will explain it in pictures.

$$A = \begin{bmatrix} \times & \times & \times & \times & \times & \times \\ \times & \times & \times & \times & \times & \times \\ & \times & \times & \times & \times & \times \\ & & \times & \times & \times & \times \\ & & & \times & \times & \times \\ & & & & \times & \times \end{bmatrix} \qquad U =$$

If, as usual, we start with $U = I$, then each core transformation is initialized to the identity.

Each iteration is begun with a move of type I acting on rows 1 and 2.

This does not disturb the Hessenberg form of either matrix. In U the transforming matrix is absorbed by a fusion operation.

At the very end of the iteration there is one more move of type I, which acts on columns $n-1$ and n.

This also does not disturb the Hessenberg form of A or U. In U, again, the transforming matrix is absorbed by fusion.

In between the two type I moves are $n-2$ moves of type II of the form $A - \lambda U \to Q_j^*(A - \lambda U)Z_{j-1}$, $j = 2, \ldots, n-1$. Because of the special way we are storing U, we deviate from the procedures that we described in Chapter 2. At the jth step we need to swap the eigenvalues of the subpencil

$$\begin{bmatrix} a_{j,j-1} & a_{jj} \\ & a_{j+1,j} \end{bmatrix} - \lambda \begin{bmatrix} u_{j,j-1} & u_{jj} \\ & u_{j+1,j} \end{bmatrix},$$

so we need to figure out the values of a few entries of U. Fortunately this is easy. If the active part of the jth core transformation is $\begin{bmatrix} c_j & -s_j \\ s_j & \bar{c}_j \end{bmatrix}$, then $u_{j,j-1} = s_{j-1}$ and $u_{jj} = \bar{c}_{j-1} c_j$, as the reader can easily check. Thus our pencil is

$$\begin{bmatrix} a_{j,j-1} & a_{jj} \\ & a_{j+1,j} \end{bmatrix} - \lambda \begin{bmatrix} s_{j-1} & \bar{c}_{j-1} c_j \\ & s_j \end{bmatrix}.$$

The two eigenvalues of this pencil are $\lambda_1 = a_{j,j-1}/s_{j-1}$ and $\lambda_2 = a_{j+1,j}/s_j$. Which procedure we follow depends on their relative magnitudes.

Primal method ($|\lambda_1| \geq |\lambda_2|$)

In Section 2.2 we laid out a primal method and a dual method for doing this swap. If $|\lambda_1| \geq |\lambda_2|$ we will use the primal method but with a twist. This computes Z_{j-1} first and then Q_j. Suppose we start out as in the primal method and compute Z_{j-1}. In our current notation we have to form the vector

$$y^T = \begin{bmatrix} s_j a_{j,j-1} - a_{j+1,j} s_{j-1} & s_j a_{jj} - a_{j+1,j} \bar{c}_{j-1} c_j \end{bmatrix},$$

then build a 2×2 unitary Z such that $y^T Z = \alpha e_2^T$ for some α. This Z is (the active part of) our Z_{j-1}.

Instead of computing Q_j, we immediately apply Z_{j-1} to the pencil. In the case $j = 3$, the picture looks like this:

When we apply the core Z to A on the right, it recombines columns 2 and 3, creating a bulge in the $(4, 2)$ position. In U let's do a turnover to move the extra core from the right to the left.

Now we need to apply Q_j^* on the left to return the matrices to upper Hessenberg form. How do we compute Q_j? The picture tells the story. The remaining red core must be Q_j, as the way to return U to Hessenberg form is to get rid of it by multiplying by its inverse, which we show in blue here.

5.2. The RQR algorithm

This returns U to upper Hessenberg form, and it also knocks out the bulge in A, returning it to Hessenberg form:

$$\begin{bmatrix} \times & \times & \times & \times & \times & \times \\ & \times & \times & \times & \times & \times \\ & & \times & \times & \times & \times \\ & & & \times & \times & \times \\ & & & & \times & \times & \times \\ & & & & & \times & \times \end{bmatrix}$$

This completes the swap.

This keeps U perfectly in Hessenberg form. The entry in the $(j+1, j-1)$ position of A will be slightly nonzero due to roundoff, but the error analysis in Section 2.3 guarantees that it is small enough to be ignored. In the primal method in the case $|\lambda_1| \geq |\lambda_2|$, Q must be determined using data from U, and that is what we have done.[16]

Dual method ($|\lambda_1| < |\lambda_2|$)

If $|\lambda_1| < |\lambda_2|$ we use the dual approach, which computes Q_j first, then Z_{j-1}. To get Q_j we must compute a vector

$$w = \begin{bmatrix} s_{j-1} a_{jj} - a_{j,j-1} \overline{c}_{j-1} c_j \\ s_{j-1} a_{j+1,j} - a_{j,j-1} s_j \end{bmatrix}$$

and then compute Q such that $Q^* w = \alpha e_1$. This Q is (the active part of) our desired Q_j. Now we apply Q_j^* to the pencil immediately. In the case $j = 3$ it looks like this:

$$\begin{bmatrix} \times & \times & \times & \times & \times & \times \\ & \times & \times & \times & \times & \times \\ & & \times & \times & \times & \times \\ & & & \times & \times & \times \\ & & & & \times & \times & \times \\ & & & & & \times & \times \end{bmatrix}$$

When we apply Q_j^* to A, it recombines rows 3 and 4, making a bulge in the $(4, 2)$ position. We pass Q_j^* through U by a turnover.

$$\begin{bmatrix} \times & \times & \times & \times & \times & \times \\ & \times & \times & \times & \times & \times \\ & & \times & \times & \times & \times \\ & \times & \times & \times & \times & \times \\ & & & & \times & \times & \times \\ & & & & & \times & \times \end{bmatrix}$$

We now return U to upper Hessenberg form by multiplying by the inverse of the extra core, which we mark in blue. This must be Z_{j-1}.

$$\begin{bmatrix} \times & \times & \times & \times & \times & \times \\ & \times & \times & \times & \times & \times \\ & & \times & \times & \times & \times \\ & \times & \times & \times & \times & \times \\ & & & & \times & \times & \times \\ & & & & & \times & \times \end{bmatrix}$$

---

[16] A look into the code for the turnover reveals that the Q_j produced by this method is (in exact arithmetic) exactly the same as would have been produced by the standard primal method.

When we apply Z_{j-1} to A on the right, it recombines columns 2 and 3, killing the bulge.

$$\begin{bmatrix} \times & \times & \times & \times & \times & \times \\ \times & \times & \times & \times & \times & \times \\ & & \times & \times & \times & \times \\ & & \times & \times & \times & \times \\ & & & \times & \times & \times \\ & & & & \times & \times \end{bmatrix}$$

This completes the swap.

Again the $(j+1, j-1)$ entry of A will be slightly nonzero due to roundoff, but the analysis in Section 2.3 guarantees that it is small enough to ignore. In the dual method, if $|\lambda_1| < |\lambda_2|$, Z should be computed using data from U, and that is what we've done.[17]

This completes our description of the RQR algorithm.

Performance of the algorithm

We wrote a Fortran implementation of the RQR algorithm and compared it with a comparable implementation of the QR algorithm, namely the code ZLAHQR from LAPACK. We found that our code generally required fewer total iterations to complete the job, was a bit faster than ZLAHQR, and was a bit more accurate as measured by backward error. These findings are reported in [22].

---

[17] See previous footnote.

Chapter 6
Krylov Processes II: Filtering

We now return for a second look at the Krylov processes that we introduced in Chapter 3.

6.1 ▪ Implicitly restarted Arnoldi process

The implicitly restarted Arnoldi process was invented by Sorensen [57] and is also described in [47, 70], for example. The description given here is a bit different from that in [70], but it is equivalent.

In Section 3.2 we introduced the Arnoldi process, which can be used to compute eigenvalues of a large, sparse matrix A. After m steps of the Arnoldi process, we will have computed orthonormal vectors $q_1, q_2, \ldots, q_{m+1}$ that span nested Krylov subspaces: $\mathrm{span}\{q_1, \ldots, q_j\} = \mathcal{K}_j(A, q_1)$, $j = 1, \ldots, m+1$. We will also have computed coefficients h_{ij} that can be placed into an upper Hessenberg matrix $H_{m+1,m}$, as explained in Section 3.2. Letting $Q_m = [q_1 \cdots q_m]$, and recalling (3.2.6) and (3.2.7), we have

$$AQ_m = Q_{m+1}H_{m+1,m} = Q_m H_m + q_{m+1}h_{m+1,m}e_m^T.$$

If the remainder term is zero, we have $AQ_m = Q_m H_m$, from which it follows that all m eigenvalues of H_m are eigenvalues of A. If, as usual, the remainder term is not zero, we can still hope that some of the eigenvalues of H_m approximate eigenvalues of A well. The quality of the approximations depends entirely on the quality of the starting vector q_1.

When we begin the process, we often have no idea of what would make a good starting vector, so we pick our q_1 at random, for example. After some Arnoldi steps we begin to get an idea of where the spectrum of A resides. The more steps we take, the more we learn. But there are limits to how many steps we are willing to take. Each step requires more computation than the previous one because the orthogonalization process includes one more vector. Moreover the storage requirement goes up because we have to store one more vector. For these reasons we generally take $m \ll n$.

The idea behind restarts is this: after m Arnoldi steps we have gained information that would allow us to restart the process with a better starting vector $\hat{q}_1$. Then m steps with this new starting vector will give us more information that allows us to restart with an even better vector, and so on. After several restarts we will have some really accurate approximations.

Let's think about what constitutes a better vector $\hat{q}_1$. To keep the discussion simple, assume that A has n linearly independent eigenvectors $v_1, \ldots, v_n$,[18] with associated eigenvalues $\lambda_1, \ldots,$

---

[18]"Most" matrices have this property. Equivalent statements: A is semisimple; A is similar to a diagonal matrix; the Jordan canonical form of A is a diagonal matrix.

λ_n. Then these n vectors form a basis of $\mathbb{C}^n$ so q_1 can be written as a linear combination

$$q_1 = c_1 v_1 + \cdots + c_k v_k + \cdots + c_n v_n.$$

Of course we have no idea what the coefficients (or the eigenvectors) are. Say, for example, we want the k eigenvalues of largest magnitude. Number the eigenvectors so that $v_1, \ldots, v_k$ are the eigenvectors associated with the k wanted eigenvalues. If we can compute the invariant subspace span$\{v_1, \ldots, v_k\}$, we will be able to extract those eigenvalues and associated eigenvectors.

Therefore we would like to construct

$$\hat{q}_1 = \hat{c}_1 v_1 + \cdots + \hat{c}_k v_k + \cdots + \hat{c}_n v_n$$

such that $\hat{c}_{k+1} = \hat{c}_{k+2} = \cdots = \hat{c}_n = 0$. This is a lot to ask for. A more modest objective would be to compute a new $\hat{q}_1$ for which $\hat{c}_1, \ldots, \hat{c}_k$ are greatly amplified with respect to $\hat{c}_{k+1}, \ldots, \hat{c}_n$.

Suppose we are able to compute a $\hat{q}_1$ such that $\hat{q}_1 = p(A)q_1$ for some polynomial p. Then

$$\hat{q}_1 = c_1 p(\lambda_1) v_1 + \cdots + c_k p(\lambda_k) v_k + \cdots + c_n p(\lambda_n) v_n.$$

If p is chosen so that $p(\lambda_1), \ldots, p(\lambda_k)$ are large relative to $p(\lambda_{k+1}), \ldots, p(\lambda_n)$, we will have made progress. This is what the implicitly restarted Arnoldi process tries to do.

To accomplish this we will have to take a number of Arnoldi steps that substantially exceeds k, say $m = k + j$, where $j = k$ or $j = 4k$, for example. Starting from q_1 we obtain $q_2, \ldots, q_{m+1}$ satisfying

$$AQ_m = Q_m H_m + q_{m+1} h_{m+1,m} e_m^T,$$

as explained previously. We now take $j = m - k$ filtering steps using a *shift* for each step. First we will say a few words about shift selection, then we will describe the filtering step, then we will state what it accomplishes.

Usually the shifts are taken to be some of the eigenvalues of H_m. We compute the eigenvalues by any method, for example, Francis's algorithm. Suppose they are $\mu_1, \ldots, \mu_m$, numbered so that

$$|\mu_1| \geq \cdots \geq |\mu_k| > |\mu_{k+1}| \geq \cdots \geq |\mu_m|.$$

The k eigenvalues of H_m of largest magnitude are estimates of the largest eigenvalues of A, so we want to enhance them and suppress the other $m - k$. Thus we use $\nu_1 = \mu_{k+1}$, $\nu_2 = \mu_{k+2}$, $\ldots$, $\nu_j = \mu_m$ as our shifts.

This is a good choice if we are looking for the eigenvalues of A of largest magnitude. If, on the other hand, we want to find the eigenvalues with largest (or smallest) real part, we can choose $\nu_1, \ldots, \nu_j$ to be the eigenvalues of H_m with smallest (or largest) real part, for example.

To be clear, it is not necessary to use shifts that are eigenvalues of H_m; we just want shifts $\nu_1, \ldots, \nu_j$ that are as near as possible to eigenvalues of A that we want to suppress.

Now we describe a single filtering step with shift ν_1. We have taken m Arnoldi steps, so we have

$$AQ_m = Q_m H_m + q_{m+1} h_{m+1,m} e_m^T. \tag{6.1.1}$$

Take a single step of Francis's algorithm with shift ν_1 on H_m. This gives

$$\tilde{H}_m = V_m^{-1} H_m V_m.$$

V_m is unitary and upper Hessenberg; it's a product of a descending sequence of $m - 1$ core transformations. Moreover by [70, §5.6] the first column of V_m is proportional to the first column of $H_m - \nu_1 I$:

$$V_m e_1 = \alpha(H_m - \nu_1 I) e_1. \tag{6.1.2}$$

6.1. Implicitly restarted Arnoldi process

Let $\tilde{Q}_m = Q_m V_m$. Multiply (6.1.1) by V_m on the right to get

$$A\tilde{Q}_m = \tilde{Q}_m \tilde{H}_m + q_{m+1} w^T, \tag{6.1.3}$$

where $w^T = h_{m+1,m} e_m^T V_m$. Because V_m is upper Hessenberg, only the last two elements of w^T are nonzero. Now delete the last column of equation (6.1.3) to get

$$A\tilde{Q}_{m-1} = \tilde{Q}_{m-1} \tilde{H}_{m-1} + \check{q}_m \check{h}_{m,m-1} e_{m-1}^T + q_{m+1} \beta e_{m-1}^T,$$

where $\check{q}_m$ is the mth column of $\tilde{Q}_m$, $\check{h}_{m,m-1}$ is the $(m, m-1)$ entry of $\tilde{H}_m$, and the other terms are self-explanatory. Define $\tilde{q}_m$ and $\tilde{h}_{m,m-1}$ by

$$\tilde{q}_m \tilde{h}_{m,m-1} = \check{q}_m \check{h}_{m,m-1} + q_{m+1} \beta,$$

where the scalar $\tilde{h}_{m,m-1}$ is chosen so that $\|\tilde{q}_m\| = 1$. Notice that q_{m+1} is orthogonal to $\check{q}_m$, which guarantees that $\tilde{q}_m \neq 0$. Moreover $\tilde{q}_m$ is orthogonal to $\mathcal{R}(\tilde{Q}_{m-1})$ because $\check{q}_m$ and q_{m+1} are. Summarizing, we have

$$A\tilde{Q}_{m-1} = \tilde{Q}_{m-1} \tilde{H}_{m-1} + \tilde{q}_m \tilde{h}_{m,m-1} e_{m-1}^T. \tag{6.1.4}$$

The filtering step is now complete. The result is an Arnoldi configuration of length $m-1$, with a different starting vector $\tilde{q}_1 = \tilde{Q}_{m-1} e_1$. We can easily determine the relationship between q_1 and $\tilde{q}_1$. Subtracting $\nu_1 Q_m$ from (6.1.1), we get

$$(A - \nu_1 I) Q_m = Q_m (H_m - \nu_1 I) + q_{m+1} h_{m+1,m} e_m^T.$$

If we now postmultiply this equation by e_1, the remainder term vanishes. Moreover, using (6.1.2) we get

$$(A - \nu_1 I) Q_m e_1 = Q_m (H_m - \nu_1 I) e_1 = \alpha^{-1} Q_m V_m e_1 = \alpha^{-1} \tilde{Q}_m e_1.$$

Thus q_1 and $\tilde{q}_1$ are related by

$$\tilde{q}_1 = \alpha (A - \nu_1 I) q_1. \tag{6.1.5}$$

Since the filtering step ends with an Arnoldi configuration (6.1.4), we can take another filtering step with shift ν_2 to get a new Arnoldi configuration of length $m-2$ with a new starting vector, and so on. After j filtering steps we have an Arnoldi configuration of length $m - j = k$:

$$A\hat{Q}_k = \hat{Q}_k \hat{H}_k + \hat{q}_{k+1} \hat{h}_{k+1,k} e_k^T. \tag{6.1.6}$$

The starting vector $\hat{q}_1$ is related to the original q_1 by

$$\hat{q}_1 = \gamma (A - \nu_j I) \cdots (A - \nu_2 I)(A - \nu_1 I) q_1 = p(A) q_1. \tag{6.1.7}$$

This is so because at each filtering step we pick up a new factor $A - \nu_i I$, as illustrated by (6.1.5). The polynomial p takes on small values in the regions near the shifts $\nu_1, \ldots, \nu_j$ and large values elsewhere, so it filters out any eigenvalues that are near $\nu_1, \ldots, \nu_j$ and enhances the others.

We can now restart with the new vector $\hat{q}_1$, but obviously we don't need to start from scratch. We already have $\hat{q}_1, \ldots \hat{q}_k$, so we can continue from the kth step. For this reason the restart is called *implicit*.

In the description of the filtering step we have shown how to keep track of the remainder term, but this is not strictly necessary. It plays no role in the filtering process, so we can ignore it! After j filtering steps, ignoring the remainder term, we arrive at (6.1.6), except that we will not know what $\hat{q}_{k+1}$ is. This is not a problem; we can compute it by taking one Arnoldi step. After j Arnoldi steps, we will be back up to an Arnoldi configuration of length m, and we can do another round of filtering, i.e., another implicit restart. We take as many restarts as we need to get the desired invariant subspace.

Shift and invert

In our discussion above we have focused on the problem of finding the k eigenvalues of largest magnitude. We also mentioned that we could equally well try to compute the k rightmost (or leftmost) eigenvalues. These are tasks for which the Arnoldi process is well suited, as it is good at locating eigenvalues on the periphery of the spectrum.[19] But what if we want to compute some eigenvalues in the "interior" of the spectrum. In this case the implicitly restarts will not work so well because the filtering and the Arnoldi steps are working at cross purposes. Here one should keep in mind the shift-and-invert strategy. If we want to find the eigenvalues that are closest to some target value τ, we can apply Arnoldi (with restarts) to the operator $(A - \tau I)^{-1}$, whose eigenvalues of largest magnitude correspond to the eigenvalues of A that are closest to τ. This works well, if you can afford it.

Krylov–Schur algorithm

An important variation on implicitly restarted Arnoldi is the Krylov–Schur method [62]. This is tangential to our development, so we will not describe it here.

Generalized Arnoldi

In Section 3.3 we introduced the generalized Arnoldi method, which replaces the continuation vector q_j by a more flexible expression (3.3.1)

$$\tilde{q}_j = \sum_{i=1}^{j} q_i k_{ij}, \quad k_{jj} \neq 0.$$

This leads to a more general Arnoldi configuration,

$$AQ_m K_m = Q_m H_m + q_{m+1} h_{m+1,m} e_m^T,$$

where (H_m, K_m) is a Hessenberg-triangular pair. The eigenvalues of (H_m, K_m) are estimates of eigenvalues of A. While this is a bit more complicated than a standard Arnoldi configuration, implicit restarts can be done in much the same way. We leave the details as an exercise for the reader.

Exercises 6.1

1. Describe a filtering step for the generalized Arnoldi process.

2. Show that (6.1.5) continues to hold in this setting, and therefore so does (6.1.7).

6.2 • Filtering the rational Krylov process

The rational Krylov process, which was introduced in Section 3.4, can be filtered and implicitly restarted in the same way as the Arnoldi process. Our presentation will look a bit different, but it amounts to the same thing. The primary sources are [13, 20, 24, 58]. See also [27].

We consider the same scenario as before. We want to find k eigenvalues in some region of interest. We take $m = k + j$ steps of the rational Krylov process, then pick j shifts $\nu_1, \ldots, \nu_j$ in regions of the plane that we wish to suppress, then we take j filtering steps using these shifts.

---
[19]But see also [43, 44].

6.2. Filtering the rational Krylov process

After m steps of the rational Krylov process we will have produced $m+1$ orthonormal vectors $q_1, \ldots, q_{m+1}$, and an equation (3.4.5)

$$AQ_{m+1}K_{m+1,m} = Q_{m+1}H_{m+1,m} \qquad (6.2.1)$$

holds, where the columns of Q_{m+1} are $q_1, \ldots, q_{m+1}$. The poles $\sigma_1, \ldots, \sigma_m$ from the m rational Krylov steps are encoded in the subdiagonal entries of the pair $(H_{m+1,m}, K_{m+1,m})$: $\sigma_i = h_{i+1,i}/k_{i+1,i}$, $i = 1, \ldots, m$. The eigenvalues of the square pencil (H_m, K_m) are estimates of the eigenvalues of A. One way to choose shifts for the filtering steps is to compute these m eigenvalues, designate the k most favorable ones as the ones we want to enhance, and use the other j, the ones we want to suppress, as the shifts $\nu_1, \ldots, \nu_j$.

Now let's describe a filtering step with shift ν_1. This is almost the same as a step of the basic RQZ algorithm applied to the nonsquare pair $(H_{m+1,m}, K_{m+1,m})$: By a move of type I, insert ν_1 at the top of the pencil, replacing pole σ_1. Then, by moves of type II, interchange pole ν_1 with σ_2, then σ_3, and so on until ν_1 reaches the bottom. The pencil $H_{m+1,m} - \lambda K_{m+1,m}$ has been transformed to an equivalent pencil

$$\tilde{H}_{m+1,m} - \lambda \tilde{K}_{m+1,m} = V^{-1}(H_{m+1,m} - \lambda K_{m+1,m})W. \qquad (6.2.2)$$

The unitary matrix V is of dimension $m+1$, it is the product of a descending sequence of core transformations, and it satisfies

$$Ve_1 = \gamma(H_{m+1,m} - \nu_1 K_{m+1,m})e_1 \qquad (6.2.3)$$

for some nonzero scalar γ. This follows from (4.1.1). See also Exercise 1.

Let

$$\tilde{Q}_{m+1} = Q_{m+1}V.$$

Then, making appropriate substitutions in (6.2.1), we get

$$A\tilde{Q}_{m+1}\tilde{K}_{m+1,m} = \tilde{Q}_{m+1}\tilde{H}_{m+1,m}. \qquad (6.2.4)$$

The filtering step is completed by deleting a row and a column. First, if we delete the last column from (6.2.4), we get

$$A\tilde{Q}_{m+1}\tilde{K}_{m+1,m-1} = \tilde{Q}_{m+1}\tilde{H}_{m+1,m-1}.$$

The last row of each of the $(m+1) \times (m-1)$ matrices $\tilde{K}_{m+1,m-1}$ and $\tilde{H}_{m+1,m-1}$ is identically zero. Therefore we can omit this row, along with the last column of $\tilde{Q}_{m+1}$, to obtain

$$A\tilde{Q}_m\tilde{K}_{m,m-1} = \tilde{Q}_m\tilde{H}_{m,m-1}. \qquad (6.2.5)$$

We have removed the pole ν_1. The poles of $(\tilde{H}_{m,m-1}, \tilde{K}_{m,m-1})$ are $\sigma_2, \ldots, \sigma_m$. This completes the filtering step.

We have replaced the original starting vector $q_1 = Q_{m+1}e_1$ by a new starting vector $\tilde{q}_1 = \tilde{Q}_m e_1$. We can discover the relationship between them by using the identity

$$(A - \nu_1 I)Q_{m+1}(H_{m+1,m} - \sigma_1 K_{m+1,m}) = (A - \sigma_1 I)Q_{m+1}(H_{m+1,m} - \nu_1 K_{m+1,m}), \quad (6.2.6)$$

which the reader can verify (Exercise 2). First note that

$$(H_{m+1,m} - \sigma_1 K_{m+1,m})e_1 = \beta e_1$$

because σ_1 is the first pole of $(H_{m+1,m}, K_{m+1,m})$. Taking this and (6.2.3) into account, equation (6.2.6) implies that

$$\beta(A - \nu_1 I)Q_{m+1}e_1 = \gamma^{-1}(A - \sigma_1 I)Q_{m+1}Ve_1,$$

so
$$\tilde{q}_1 = Q_{m+1} V e_1 = \alpha (A - \sigma_1 I)^{-1}(A - \nu_1 I) q_1,$$
where $\alpha = \gamma \beta$. Thus the filter is the rational function $(\lambda - \nu_1)/(\lambda - \sigma_1)$. This suppresses eigenvalues near ν_1 and amplifies eigenvalues near σ_1.

We have described one filtering step. After this we can do a second filtering step with shift ν_2 starting from (6.2.5). This removes the pole σ_2 and applies a rational filter $(\lambda - \nu_2)/(\lambda - \sigma_2)$. After $j - 2$ additional filtering steps we arrive at

$$A \hat{Q}_{k+1} \hat{K}_{k+1,k} = \hat{Q}_{k+1} \hat{H}_{k+1,k} \tag{6.2.7}$$

with
$$\hat{q}_1 = r(A) q_1,$$
where
$$r(\lambda) = \prod_{i=1}^{j} \frac{\lambda - \nu_i}{\lambda - \sigma_i}.$$

Eigenvalues of A near $\nu_1, \ldots, \nu_j$ are suppressed, while those near $\sigma_1, \ldots, \sigma_j$ are amplified. The remaining poles in (6.2.7) are $\sigma_{j+1}, \ldots, \sigma_m$.[20]

After the filtering steps we can restart with the vector $\hat{q}_1$. Of course we do not start from scratch, as the first k steps are determined by (6.2.7). Again we have an *implicit* restart.

Exercises 6.2

1. Review the moves of types I and II in Section 4.1. Verify that the matrix V in (6.2.2) is the product of a descending sequence of m core transformations. In addition, show that W is the product of a descending sequence of $m - 1$ core transformations. Show that V satisfies (6.2.3).

2. Prove that if $AQ_{m+1} K_{m+1,m} = Q_{m+1} H_{m+1,m}$, then
$$(A - \nu I) Q_{m+1}(H_{m+1,m} - \sigma K_{m+1,m}) = (A - \sigma I) Q_{m+1}(H_{m+1,m} - \nu K_{m+1,m})$$
for any scalars ν and σ. One way to do this is to multiply out the two sides of the purported equation and compare them.

---

[20] The order of the poles is not set in stone. Before the filtering steps are begun (or at any intermediate point), the poles $\sigma_1, \ldots, \sigma_m$ can be reordered as needed by moves of type II.

Chapter 7
Block Algorithms

In Chapter 4 we developed algorithms that swap poles one pair at a time. Swapping a pair of poles is the same as swapping a pair of eigenvalues of the pole pencil, which can be done by methods described in Chapter 2. In that chapter we also showed how to swap whole blocks of eigenvalues, and in this chapter we consider algorithms that swap the poles in blocks. A primary reference is [58].

An obvious motivation for this is the case of real pencils. Some of the eigenvalues may be complex, but these occur in complex conjugate pairs. To find complex eigenvalues we will need to use complex shifts, but they can also be taken in conjugate pairs, which can be housed in real 2×2 blocks. This allows for an algorithm that stays entirely in real arithmetic by swapping 2×2 blocks.

7.1 ▪ An algorithm that swaps blocks

Although the 2×2 case is our primary motivation, it is possible in principle to swap blocks of any size. Consider a block Hessenberg pencil $A - \lambda B$, where

$$A = \begin{bmatrix} * & * & \cdots & * & * \\ A_{21} & A_{22} & \cdots & A_{2,m-1} & * \\ & A_{32} & \cdots & A_{3,m-1} & * \\ & & \ddots & \vdots & \vdots \\ & & & A_{m,m-1} & * \end{bmatrix}, \qquad (7.1.1)$$

and B is partitioned in conformity with A. The asterisks denote the first row and last column. Thus the pole pencil is

$$A_\pi - \lambda B_\pi = \begin{bmatrix} A_{21} & A_{22} & \cdots & A_{2,m-1} \\ & A_{32} & \cdots & A_{3,m-1} \\ & & \ddots & \vdots \\ & & & A_{m,m-1} \end{bmatrix} - \lambda \begin{bmatrix} B_{21} & B_{22} & \cdots & B_{2,m-1} \\ & B_{32} & \cdots & B_{3,m-1} \\ & & \ddots & \vdots \\ & & & B_{m,m-1} \end{bmatrix}. \qquad (7.1.2)$$

The diagonal blocks $A_{k+1,k} - \lambda B_{k+1,k}$ are square of varying sizes, and their eigenvalues are the poles of (A, B). From Chapter 2 we know how to swap any two adjacent blocks. These are our moves of type II. Now all we need is to develop moves of type I, i.e., moves that inject new poles into the first and last blocks.

Theorem 1.2.1 (generalized Schur) tells us that each of the diagonal blocks is unitarily equivalent to a triangular pencil:

$$U^*_{k+1}(A_{k+1,k} - \lambda B_{k+1,k})V_k = \check{A}_{k+1,k} - \lambda \check{B}_{k+1,k},$$

where $\check{A}_{k+1,k}$ and $\check{B}_{k+1,k}$ are triangular matrices. Now define unitary block-diagonal matrices $U, V \in \mathbb{C}^{n \times n}$ by

$$U = \text{diag}\{1, U_2, U_3, \ldots, U_m\} \quad \text{and} \quad V = \text{diag}\{V_1, V_2, \ldots, V_{m-1}, 1\}. \qquad (7.1.3)$$

Then

$$U^*(A - \lambda B)V = \check{A} - \lambda \check{B}, \qquad (7.1.4)$$

where $\check{A} - \lambda \check{B}$ is a Hessenberg pencil. In summary, every block Hessenberg pencil is unitarily equivalent to a true Hessenberg pencil via block-diagonal unitary transformations of the form (7.1.3), which do not alter the eigenvalues of the subdiagonal blocks. These transformations are far from unique. The eigenvalues of each block $A_{k+1,k} - \lambda B_{k+1,k}$ can be made to appear in any order on the main diagonal of $\check{A}_{k+1,k} - \lambda \check{B}_{k+1,k}$. We call $A - \lambda B$ a *proper* block Hessenberg pencil if there is an ordering for which $\check{A} - \lambda \check{B}$ is proper. If $[\sigma_1, \ldots, \sigma_{n-1}]$ is the ordered pole set of $\check{A} - \lambda \check{B}$, we will also call it an ordered pole set (not unique) of $A - \lambda B$.

Move of type I at the top

Recall from Section 4.7 the notation

$$M(\rho, \sigma) = (A - \rho B)(A - \sigma B)^{-1}.$$

Suppose the size of the first block $A_{21} - \lambda B_{21}$ is $s \times s$ and its eigenvalues are $\sigma_1, \ldots, \sigma_s$. These are the first s poles of $A - \lambda B$. We want to replace $\sigma_1, \ldots, \sigma_s$ by s new poles (shifts) ρ_1, $\ldots, \rho_s$, all distinct from $\sigma_1, \ldots, \sigma_s$. First we compute

$$x = M(\rho_s, \sigma_s) \cdots M(\rho_2, \sigma_2) M(\rho_1, \sigma_1) e_1. \qquad (7.1.5)$$

We will show below that $x \in \mathcal{E}_{s+1} = \text{span}\{e_1, \ldots, e_{s+1}\}$. Therefore there is a unitary matrix

$$Q = \begin{bmatrix} Q_{s+1} & \\ & I \end{bmatrix} \in \mathbb{C}^{n \times n} \quad \text{with} \quad Q_{s+1} \in \mathbb{C}^{(s+1) \times (s+1)} \qquad (7.1.6)$$

such that $Q^*x = \alpha e_1$, where $|\alpha| = \|x\|$. Now let $\tilde{A} - \lambda \tilde{B} = Q^*(A - \lambda B)$. This transformation does not alter the block structure of the pencil, as it only changes the first $s+1$ rows. It changes $A_{21} - \lambda B_{21}$ to, say, $\tilde{A}_{21} - \lambda \tilde{B}_{21}$ having poles $\rho_1, \rho_2, \ldots, \rho_s$. Thus this transformation by Q^* is the desired move of type I. We now prove these claims.

Proof. Referring back to (7.1.4), let $\check{M}(\rho, \sigma) = (\check{A} - \rho \check{B})(\check{A} - \sigma \check{B})^{-1}$. The reader can easily check that $M(\rho, \sigma) = U\check{M}(\rho, \sigma)U^*$, so

$$x = U\check{M}(\rho_s, \sigma_s) \cdots \check{M}(\rho_1, \sigma_1) U^* e_1.$$

Since $U^*e_1 = e_1$, we can forget about U^*. Applying Proposition 4.7.6 to $(\check{A}, \check{B})$, we see that $\mathcal{K}_{s+1}(\check{A}, \check{B}, e_1, [\sigma_1, \ldots, \sigma_s]) = \mathcal{E}_{s+1}$. Since $\check{M}(\rho_s, \sigma_s) \cdots \check{M}(\rho_1, \sigma_1)e_1$ is in this rational Krylov subspace, it is in $\mathcal{E}_{s+1}$. Since $U\mathcal{E}_{s+1} = \mathcal{E}_{s+1}$, we see that $x \in \mathcal{E}_{s+1}$.

Since the factors $M(\rho_i, \sigma_i)$ that define x commute with one another, we can reorder them however we please. Thus for any j we can write

$$x = M(\rho_j, \sigma_j) \left(\prod_{i \neq j} M(\rho_i, \sigma_i) \right) e_1 = (A - \rho_j B)(A - \sigma_j B)^{-1} \left(\prod_{i \neq j} M(\rho_i, \sigma_i) \right) e_1.$$

7.1. An algorithm that swaps blocks

Let
$$y = (A - \sigma_j B)^{-1} \left(\prod_{i \neq j} M(\rho_i, \sigma_i) \right) e_1, \tag{7.1.7}$$

so that $x = (A - \rho_j B)y$. It is not hard to show that $y \in \mathcal{E}_{s+1}$ (Exercise 1). In fact $y \in \mathcal{E}_s$. If not, then $y_{s+1} \neq 0$, so $x_{s+2} = (a_{s+2,s+1} - \rho_j b_{s+2,s+1})y_{s+1} \neq 0$ because ρ_j is not an eigenvalue of $A_{21} - \lambda B_{21}$. Thus $y_{s+1} = 0$. Let $\hat{x} \in \mathbb{C}^{s+1}$ and $\hat{y} \in \mathbb{C}^s$ denote the nonzero parts of x and y, respectively. Let $\widehat{A} - \lambda \widehat{B}$ denote the $(s+1) \times s$ subpencil in the upper-left-hand corner of $A - \lambda B$. This consists of part of the first row of $A - \lambda B$ atop the block $A_{21} - \lambda B_{21}$.

Then $\hat{x} = (\widehat{A} - \rho_j \widehat{B})\hat{y}$. Now build an $(s+1) \times (s+1)$ *initial bulge pencil* [67]
$$A_0 - \lambda B_0 = \left[\, \hat{x} \mid \widehat{A} \,\right] - \lambda \left[\, 0 \mid \widehat{B} \,\right].$$

This pencil has an infinite eigenvalue, as B_0 is singular. The other s eigenvalues turn out to be $\rho_1, \ldots, \rho_s$. This is so because $\begin{bmatrix} -1 \\ \hat{y} \end{bmatrix}$ is an eigenvector associated with ρ_j. Since j could be any of 1, 2, ..., s, we see that $\rho_1, \ldots, \rho_s$ are all eigenvalues. This gives the complete spectrum of $A_0 - \lambda B_0$ if $\rho_1, \ldots, \rho_s$ are distinct. This is also true by continuity if some of the ρ_j are repeated.

If we apply Q^*_{s+1} from (7.1.6) to $A_0 - \lambda B_0$, we get
$$Q^*_{s+1}(A_0 - \lambda B_0) = \left[\begin{array}{c|ccc} \alpha & * & \cdots & * \\ \hline 0 & \multicolumn{3}{c}{\widetilde{A}_{21} - \lambda \widetilde{B}_{21}} \end{array}\right].$$

We see the infinite eigenvalue deflated in the upper-left-hand corner, so the eigenvalues $\rho_1, \ldots, \rho_s$ must reside in $\widetilde{A}_{21} - \lambda \widetilde{B}_{21}$. This is the desired result. □

Move of type I at the bottom

Injecting poles at the bottom is about the same as at the top. Recall from Section 4.7 the notation
$$N(\rho, \sigma) = (A - \sigma B)^{-1}(A - \rho B).$$

Suppose the bottom block $A_{m,m-1} - \lambda B_{m,m-1}$ is $s \times s$, it has poles $\rho_1, \ldots, \rho_s$, and we want to replace them by $\tau_1, \ldots, \tau_s$. Compute
$$w^T = e_n^T N(\rho_1, \tau_1) N(\rho_2, \tau_2) \cdots N(\rho_s, \tau_s). \tag{7.1.8}$$

Then only the last $s+1$ entries of w^T can be nonzero, so there is a unitary Z of the form
$$Z = \begin{bmatrix} I & \\ & Z_{s+1} \end{bmatrix}$$

such that $w^T Z = \alpha e_n^T$. The transformation
$$(A - \lambda B)Z = \widetilde{A} - \lambda \widetilde{B}$$

inserts shifts $\tau_1, \ldots, \tau_s$ into $\widetilde{A}_{m+1,m} - \lambda \widetilde{B}_{m+1,m}$. The details are left as an exercise for the reader.

Summary

Now that we know how to do moves of types I and II, we can use them to build block generalizations of the various algorithms presented in Chapter 4.

A comprehensive treatment of the block algorithms would include a discussion of how to compute the vectors x in (7.1.5) and w^T in (7.1.8) in an efficient way. We leave this as an exercise for the reader. In the next section we will show how to do it in the special case $s = 2$.

We could also address the question of convergence theorems for block algorithms, but we won't. In the next section we will present a convergence result for the special case of 2×2 blocks.

Exercises 7.1

1. Show that the vector y defined by (7.1.7) satisfies $y \in \mathcal{E}_{s+1}$. Hints: (1) Make use of the transformation (7.1.4), bearing in mind that the poles $\sigma_1, \ldots, \sigma_s$ can be made to appear in any order in $\check{A}_{21} - \lambda \check{B}_{21}$. In particular, σ_j can be last. (2) Proceed as in the proof of Proposition 4.7.6.

2. Prove the claims about the move of type I at the bottom. In particular, show that only the last $s + 1$ entries of y^T can be nonzero. For this you might imitate the argument from Proposition 4.7.6.

7.2 ▪ The Case of 2×2 Blocks

Let's take a closer look at the real 2×2 case. Suppose our pair (A, B) is real and block Hessenberg, with the blocks no larger than 2×2. A typical 2×2 block will carry a complex conjugate pair of poles, but it could as well have two real poles. Typically there will also be some 1×1 blocks carrying single real poles. We showed how to swap blocks of these (and all) sizes in Chapter 2. The reader can easily check that since all the matrix entries are real, the swaps proceed entirely in real arithmetic and yield real results. Since we are swapping 2×2 blocks, refinement steps will sometimes be necessary.

Now let's look at moves of type I in the 2×2 case. The first block $A_{21} - \lambda B_{21}$ is 2×2; suppose its eigenvalues are σ_1 and σ_2. These are the first two poles of $A - \lambda B$. We want to replace them by two new poles (shifts) ρ_1 and ρ_2. These could be real or a complex conjugate pair. Recall from (7.1.5) that we begin by computing

$$x = M(\rho_2, \sigma_2) M(\rho_1, \sigma_1) e_1$$
$$= (A - \rho_2 B)(A - \sigma_2 B)^{-1}(A - \rho_1 B)(A - \sigma_1 B)^{-1} e_1. \qquad (7.2.1)$$

From the discussion in Section 7.1 we know that $x \in \mathcal{E}_3$; only the first three entries of x can be nonzero, and we need a method to compute these three numbers with minimum effort. Here's one way to do it.

Referring back to the discussion leading up to (7.1.4), we know that $A - \lambda B$ is unitarily equivalent to a (complex) Hessenberg pencil $\check{A} - \lambda \check{B}$, and

$$x = U(\check{A} - \rho_2 \check{B})(\check{A} - \sigma_2 \check{B})^{-1}(\check{A} - \rho_1 \check{B})(\check{A} - \sigma_1 \check{B})^{-1} U^* e_1. \qquad (7.2.2)$$

We will assume that $(\check{A}, \check{B})$ is a proper Hessenberg pair, since otherwise we can reduce the problem immediately. Our computation will not require us to compute the entire pencil $\check{A} - \lambda \check{B}$; we will only need the first block $\check{A}_{21} - \lambda \check{B}_{21} = U_2^*(A_{21} - \lambda B_{21})V_1$. This transformation can be achieved by a 2×2 generalized Schur decomposition, which costs almost nothing. If the poles σ_1 and σ_2 are complex conjugates, $\check{A}_{21}$ and $\check{B}_{21}$ will be complex. They are upper triangular, and the eigenvalues of $(\check{A}_{21}, \check{B}_{21})$ are σ_1 and σ_2. Let's suppose we have numbered them so that σ_1 is in the $(1, 1)$ position.

7.2. The Case of 2×2 Blocks

In (7.2.2) $U^* e_1 = e_1$, so we can ignore U^*. Let

$$y = (\check{A} - \sigma_1 \check{B})^{-1} e_1, \quad z = (\check{A} - \rho_1 \check{B})y, \quad w = (\check{A} - \sigma_2 \check{B})^{-1} z, \quad v = (\check{A} - \rho_2 \check{B})w,$$

so that $x = Uv$.

For the computation of y we need to solve the linear system

$$(\check{A} - \sigma_1 \check{B})y = e_1.$$

The $(2, 1)$ entry of the matrix is zero: $\check{a}_{21} - \sigma_1 \check{b}_{21} = 0$, because σ_1 is the first pole. (To understand this recall the relationship of the subpencil $\check{A}_{21} - \lambda \check{B}_{21}$ to the larger pencil $\check{A} - \lambda \check{B}$.) It follows that $y = \alpha e_1$ for some α (Exercise 1). The scalar α can be ignored because any multiple of x can be used in place of x itself. This means that the y computation can be skipped altogether. Next we have

$$z = \alpha(\check{A} - \rho_1 \check{B})e_1.$$

Ignoring α, we see that z is just the first column of $\check{A} - \rho_1 \check{B}$. This matrix is upper Hessenberg, so $z \in \mathcal{E}_2$. Now we compute w by solving

$$(\check{A} - \sigma_2 \check{B})w = z.$$

Because σ_2 is the second pole, we have $\check{a}_{32} - \sigma_2 \check{b}_{32} = 0$. Thus the system is block triangular and, since $z \in \mathcal{E}_2$, we also have $w \in \mathcal{E}_2$, and we can compute w by solving a 2×2 system (Exercise 2). Next we compute v by

$$v = (\check{A} - \rho_2 \check{B})w.$$

Since $w \in \mathcal{E}_2$ and the matrix is upper Hessenberg, we have $v \in \mathcal{E}_3$. The computation of v requires only the first two columns of $\check{A} - \rho_2 \check{B}$. Finally we have

$$x = Uv.$$

The block-diagonal structure of U implies that $x \in \mathcal{E}_3$.

The reader can easily check that the computation of x requires knowledge of only U_2, V_1, and the first two columns of $\check{A}$ and $\check{B}$, and requires $O(1)$ arithmetic. The arithmetic is complex, but the resulting x is real (Exercise 3).

Now consider a move of type I at the bottom of the pencil. Suppose the bottom block $A_{m,m-1} - \lambda B_{m,m-1}$ is 2×2 with poles ρ_1 and ρ_2, and we want to replace those with poles τ_1 and τ_2. This can be done by a procedure entirely analogous to the move of type I at the top. We leave the details for the reader.

Shift selection

The most obvious strategy is to compute the eigenvalues of the lower-right-hand 2×2 subpencil and introduce them as shifts at the top and chase them to the bottom. We can then compute the eigenvalues of the upper-left-hand 2×2 subpencil and introduce them as poles at the bottom at the end of the iteration. This strategy is not foolproof. Exceptional shifts can be introduced in the rare situations when many iterations pass with no eigenvalues converging. One can imagine many variants of this scheme.

If the pencil is large enough, aggressive early deflation can be used. This can be done at both bottom and top to provide streams of shifts to be introduced at the top and poles to be introduced at the bottom.

Numerical experiments

Steel et al. [58] wrote a real, double-shift code that employs aggressive early deflation at both the bottom and top of the pencil. Their tests demonstrate that their code is consistently faster than a comparable QZ code on a variety of problems from various applications, due to fewer iterations being required.

A convergence theorem

Here we will demonstrate the effect of swapping two adjacent two 2×2 blocks. The argument that we present here can be readily generalized to the swap of any two blocks of the same size $s \times s$, where s can be anything.

We start by proving a key lemma, which we present in full generality. Thus, consider a block Hessenberg pencil with blocks of any size, as in (7.1.1).

Lemma 7.2.1. *Suppose the square blocks $A_{k+1,k} - \lambda B_{k+1,k}$ in (7.1.2) are $s_k \times s_k$ for $k = 1$, ..., $m - 1$. Then, for $j = 1 + s_1, 1 + s_1 + s_2, \ldots, 1 + s_1 + \cdots + s_{m-1}$,*

$$\mathcal{K}_j(A, B, e_1, [\sigma_1, \ldots, \sigma_{j-1}]) = \mathcal{E}_j.$$

Proof. We will use the transformation $U^*(A - \lambda B)V = \check{A} - \lambda \check{B}$ from (7.1.4). Due to the sizes of the blocks, we have $U^* \mathcal{E}_j = \mathcal{E}_j$ for the allowed values of j. Moreover $U^* e_1 = e_1$. Since $\check{A} - \lambda \check{B}$ is a Hessenberg pencil, Proposition 4.7.6 implies that $\mathcal{K}_j(\check{A}, \check{B}, e_1, [\sigma_1, \ldots, \sigma_{j-1}]) = \mathcal{E}_j$. Therefore

$$\begin{aligned}\mathcal{K}_j(A, B, e_1, [\sigma_1, \ldots, \sigma_{j-1}]) &= U \mathcal{K}_j(\check{A}, \check{B}, U^* e_1, [\sigma_1, \ldots, \sigma_{j-1}]) \\ &= U \mathcal{K}_j(\check{A}, \check{B}, e_1, [\sigma_1, \ldots, \sigma_{j-1}]) \\ &= U \mathcal{E}_j = \mathcal{E}_j.\end{aligned}$$ □

Remark 7.2.2. *We would like to make a similar statement about the $\mathcal{L}_{j-1}$ spaces, but the argument we have used here does not carry over because $V e_1 \neq e_1$, among other problems.*

Now let's consider what happens when we swap two 2×2 blocks. Suppose we swap a block containing σ_{j-2} and σ_{j-1} with a block containing σ_j and σ_{j+1}, where $3 \leq j \leq n - 2$. The blocks occupy rows $j - 1$ to $j + 2$ and columns $j - 2$ to $j + 1$. The transformation is

$$\hat{A} - \lambda \hat{B} = Q^*(A - \lambda B)Z. \tag{7.2.3}$$

Theorem 7.2.3. *The double pole swap described directly above effects the subspace transformations*

$$\mathcal{Q}_j = Q\mathcal{E}_j = M(\sigma_{j-2}, \sigma_j) M(\sigma_{j-1}, \sigma_{j+1}) \mathcal{E}_j$$

and

$$\mathcal{Z}_{j-1} = Z\mathcal{E}_{j-1} = N(\sigma_{j-2}, \sigma_j) N(\sigma_{j-1}, \sigma_{j+1}) \mathcal{E}_{j-1}.$$

The change of coordinate system (7.2.3) maps $\mathcal{Q}_j$ back to $\mathcal{E}_j$ and $\mathcal{Z}_{j-1}$ back to $\mathcal{E}_{j-1}$.

Remark 7.2.4. *Recall that*

$$M(\sigma_{j-2}, \sigma_j) M(\sigma_{j-1}, \sigma_{j+1}) = (A - \sigma_{j-2}B)(A - \sigma_j B)^{-1}(A - \sigma_{j-1}B)(A - \sigma_{j+1}B)^{-1}.$$

7.2. The Case of 2×2 Blocks

The poles σ_{j-2} and σ_{j-1}, which are moving downward, are associated with positive powers, while σ_j and σ_{j+1} are associated with negative powers. The exact order of the poles in these expressions is not rigidly determined. For example,

$$M(\sigma_{j-2}, \sigma_j) M(\sigma_{j-1}, \sigma_{j+1}) = M(\sigma_{j-1}, \sigma_j) M(\sigma_{j-2}, \sigma_{j+1}),$$

so σ_{j-2} is not rigidly paired with σ_j, nor is σ_{j-1} rigidly paired with σ_{j+1}.

Remark 7.2.5. *We will prove the $\mathcal{Q}_j$ assertion, from which we can deduce the $\mathcal{Z}_{j-1}$ assertion by an argument using pertransposes and orthogonal complements. See Exercise 7.*

Proof. Lemma 7.2.1 can be applied to both (A, B) and $(\hat{A}, \hat{B})$. An ordered pole set of $(\hat{A}, \hat{B})$ is

$$[\sigma_1, \ldots, \sigma_{j-3}, \sigma_j, \sigma_{j+1}, \sigma_{j-2}, \sigma_{j-1}, \sigma_{j+2}, \ldots, \sigma_{n-1}],$$

so

$$\mathcal{E}_j = \mathcal{K}_j(\hat{A}, \hat{B}, e_1, [\sigma_1, \ldots, \sigma_{j-3}, \sigma_j, \sigma_{j+1}]).$$

Therefore

$$\mathcal{Q}_j = Q\mathcal{E}_j = Q\mathcal{K}_j(\hat{A}, \hat{B}, e_1, [\sigma_1, \ldots, \sigma_{j-3}, \sigma_j, \sigma_{j+1}])$$
$$= \mathcal{K}_j(A, B, Qe_1, [\sigma_1, \ldots, \sigma_{j-3}, \sigma_j, \sigma_{j+1}]).$$

Noting that $Qe_1 = e_1$, and using part (a) of Theorem 4.7.4 twice, we obtain

$$\mathcal{Q}_j = \mathcal{K}_j(A, B, e_1, [\sigma_1, \ldots, \sigma_{j-3}, \sigma_j, \sigma_{j+1}])$$
$$= M(\rho, \sigma_j) M(\rho, \sigma_{j+1}) \left(\prod_{i=1}^{j-3} M(\rho, \sigma_i) \right) \mathcal{K}_j(M(\hat{\rho}, \rho), e_1)$$
$$= M(\sigma_{j-2}, \rho) M(\sigma_{j-1}, \rho) M(\rho, \sigma_j) M(\rho, \sigma_{j+1}) \left(\prod_{i=1}^{j-1} M(\rho, \sigma_i) \right) \mathcal{K}_j(M(\hat{\rho}, \rho), e_1)$$
$$= M(\sigma_{j-2}, \sigma_j) M(\sigma_{j-1}, \sigma_{j+1})) \mathcal{K}_j(A, B, e_1, [\sigma_1, \ldots, \sigma_{j-2}, \sigma_{j-1}])$$
$$= M(\sigma_{j-2}, \sigma_j) M(\sigma_{j-1}, \sigma_{j+1})) \mathcal{E}_j,$$

which is the desired result. Here we have used facts like $M(\rho, \sigma_{j-2}) = M(\sigma_{j-2}, \rho)^{-1}$ and $M(\sigma_{j-2}, \rho) M(\rho, \sigma_j) = M(\sigma_{j-2}, \sigma_j)$, and in the final step we used Lemma 7.2.1 once more. $\square$

Now let's generalize this result to two $s \times s$ blocks. Suppose we swap a block containing $\sigma_{j-s}, \ldots, \sigma_{j-1}$ with a block containing $\sigma_j, \ldots, \sigma_{j+s-1}$. The blocks occupy rows $j-s+1$ to $j+s$ and columns $j-s$ to $j+s-1$. The transformation is

$$\hat{A} - \lambda \hat{B} = Q^*(A - \lambda B)Z. \tag{7.2.4}$$

Theorem 7.2.6. *The s-fold pole swap described directly above effects the subspace transformations*

$$\mathcal{Q}_j = Q\mathcal{E}_j = \prod_{i=0}^{s-1} M(\sigma_{j-s+i}, \sigma_{j+i}) \mathcal{E}_j$$

and

$$\mathcal{Z}_{j-1} = Z\mathcal{E}_{j-1} = \prod_{i=0}^{s-1} N(\sigma_{j-s+i}, \sigma_{j+i}) \mathcal{E}_{j-1}.$$

The change of coordinate system (7.2.4) maps $\mathcal{Q}_j$ back to $\mathcal{E}_j$ and $\mathcal{Z}_{j-1}$ back to $\mathcal{E}_{j-1}$.

The proof is a straightforward extension of that of Theorem 7.2.3; the reader is invited to fill in the details.

In the following section we will warn the reader that it might not be a good idea to swap very large blocks because of potential problems with roundoff errors. The theorem is, nevertheless, useful even for large s, for the following reason. Suppose we have a bunch of smallish blocks that we are transforming downward, and the sum of their dimensions is s. We can treat these as one large $s \times s$ block. Suppose that directly below that we have another bunch of smallish blocks that we are transforming upward, and the sum of their dimensions is also s. We can treat that as another $s \times s$ block. Theorem 7.2.6 tells us the effect of swapping the two $s \times s$ blocks. But in practice we do not need to swap the two large blocks all at once. Instead we can swap the small blocks in pairs until all of the small blocks that were in the upper $s \times s$ block have been moved to the lower $s \times s$ block, and vice versa. Theorem 7.2.6 remains relevant; it tells the cumulative effect of all the small block swaps.

Exercises 7.2

1. Suppose $\check{A} - \lambda \check{B}$ is a proper Hessenberg pencil with ordered pole set
$$[\sigma_1, \sigma_2, \ldots, \sigma_{n-1}],$$
none of which is an eigenvalue of the pencil.

    (a) Show that $\check{a}_{21} - \sigma_1 \check{b}_{21} = 0$ (so the matrix $\check{A} - \sigma_1 \check{B}$ is block triangular).
    (b) Show that $\check{a}_{11} - \sigma_1 \check{b}_{11} \neq 0$.
    (c) Suppose $y = \begin{bmatrix} y_1 \\ \tilde{y} \end{bmatrix}$ is the solution of $(\check{A} - \sigma_1 \check{B})y = e_1$. Show that $\tilde{y} = 0$ and $y_1 = \alpha$, where $\alpha = (\check{a}_{11} - \sigma_1 \check{b}_{11})^{-1}$. Therefore $y = \alpha e_1$.

2. Under the same assumptions as in the previous exercise, consider the linear system $(\check{A} - \sigma_2 \check{B})w = z$, where $z \in \mathcal{E}_2$.

    (a) Show that $\check{a}_{32} - \sigma_2 \check{b}_{32} = 0$.
    (b) Partition the system as
    $$\begin{bmatrix} C_{11} & C_{12} \\ & C_{22} \end{bmatrix} \begin{bmatrix} \hat{w} \\ \tilde{w} \end{bmatrix} = \begin{bmatrix} \hat{z} \\ 0 \end{bmatrix},$$
    where C_{11} is 2×2. Show that C_{11} and C_{22} are both nonsingular (easy).
    (c) Show that $\tilde{w} = 0$ and $C_{11}\hat{w} = \hat{z}$. Thus $w \in \mathcal{E}_2$, and it can be computed by solving a 2×2 system.

3. Prove that x is real. Here is one way to do it: Start with the definition (7.2.1). Assume temporarily that B^{-1} exists, and rewrite the definition of x in terms of AB^{-1}. Then use commutativity. If ρ_1 is not real, then $\rho_2 = \overline{\rho}_1$, and similarly for σ_1. Deduce that x is real. Deduce by continuity that the result also holds if B is singular.

4. Explain in detail how to do a move of type I at the bottom.

5. (Duality in subspace iteration) Let $\mathcal{S} \subseteq \mathbb{C}^n$ be any subspace and let $C \in \mathbb{C}^{n \times n}$ be any nonsingular matrix.

    (a) Suppose $s \in \mathcal{S}$ and $s^\perp \in \mathcal{S}^\perp$. Show that the vectors Cs and $(C^*)^{-1}s^\perp$ are orthogonal.
    (b) Show that if $\mathcal{U} = C\mathcal{S}$ then $\mathcal{U}^\perp = (C^*)^{-1}\mathcal{S}^\perp$.

7.2. The Case of 2×2 Blocks

6. (a) Let $F \in \mathbb{C}^{n \times n}$ be the *flip* or *anti-identity* matrix defined by $F = [e_n \ e_{n-1} \ \cdots \ e_1]$. Show that $F = F^{-1} = F^*$. Show that the operation $C \to FC$ flips the rows of C, and the operation $C \to CF$ flips the columns of C.

   (b) For any $C \in \mathbb{C}^{n \times n}$ define the *pertranspose* (or *conjugate pertranspose*) of C by $C^P = FC^*F$. This operation reverses the rows and columns of C and takes the conjugate transpose. Show that if C is upper triangular, then so is C^P. Show that the eigenvalues of C^P are the complex conjugates of the eigenvalues of C. Show that $C^{PP} = C$.

   (c) Show that if (A, B) is a block Hessenberg pair with block sizes $s_1, \ldots, s_m$ and ordered pole set $\sigma_1, \ldots, \sigma_{n-1}$, then (A^P, B^P) is a block Hessenberg pair with block sizes $s_m, \ldots, s_1$ and ordered pole set $\overline{\sigma}_{n-1}, \ldots, \overline{\sigma}_1$.

7. In this exercise we prove the $\mathcal{Z}_{j-1}$ assertion of Theorem 7.2.3. Exercises 5 and 6 are prerequisites. The details may seem daunting, but the idea is simple. Turn the pencil upside down. Get a result for the upside-down pencil. Then take orthogonal complements to get the desired result.

   (a) Referring to (7.2.3), show that
   $$\hat{A}^P - \overline{\lambda}\hat{B}^P = Z^P(A^P - \overline{\lambda}B^P)(Q^P)^*. \tag{7.2.5}$$

   (b) The two blocks discussed in Theorem 7.2.3 correspond to two 2×2 blocks in the pertransposed pencil. Show that these consist of a block containing poles $\overline{\sigma}_j$ and $\overline{\sigma}_{j+1}$ atop a block containing poles $\overline{\sigma}_{j-2}$ and $\overline{\sigma}_{j-1}$, and these occupy rows $n - j - 1$ to $n - j + 2$ and columns $n - j - 2$ to $n - j + 1$ in the pertransposed pencil. These blocks get swapped in the transformation (7.2.5).

   (c) Apply the $\mathcal{Q}_j$ part of Theorem 7.2.3 to (7.2.5) with j replaced by $n - j + 1$ to deduce that
   $$(Z^P)^* \mathcal{E}_{n-j+1} = \cdots$$
   $$(A^P - \overline{\sigma}_j B^P)(A^P - \overline{\sigma}_{j-2} B^P)^{-1}(A^P - \overline{\sigma}_{j+1} B^P)(A^P - \overline{\sigma}_{j-1} B^P)^{-1} \mathcal{E}_{n-j+1}.$$

   (d) Now take the orthogonal complement of this equation and use the duality result from Exercise 5 to obtain
   $$(Z^P)^{-1} \mathcal{E}^\perp_{n-j+1} = \cdots$$
   $$(A^P - \overline{\sigma}_j B^P)^{*-1}(A^P - \overline{\sigma}_{j-2} B^P)^*(A^P - \overline{\sigma}_{j+1} B^P)^{*-1}(A^P - \overline{\sigma}_{j-1} B^P)^* \mathcal{E}^\perp_{n-j+1}.$$

   (e) Now multiply both sides by F on the left and simplify as much as possible, noting that $F \mathcal{E}^\perp_{n-j+1} = \mathcal{E}_{j-1}$, to obtain
   $$Z\mathcal{E}_{j-1} = (A - \sigma_j B)^{-1}(A - \sigma_{j-2} B)(A - \sigma_{j+1} B)^{-1}(A - \sigma_{j-1} B)\mathcal{E}_{j-1}$$
   $$= N(\sigma_{j-2}, \sigma_j) N(\sigma_{j-1}, \sigma_{j+1}) \mathcal{E}_{j-1},$$
   which is the desired result.

8. Sketch a proof of Theorem 7.2.6.

7.3 • Bulge chasing as pole swapping

In Chapter 4, especially Section 4.2, we established a connection between bulge chasing and pole swapping. In that chapter we restricted our attention to chases of single shifts and swaps of single poles. Here we will extend our considerations to chases of multiple shifts and swaps of blocks of poles.

The bulge-chasing algorithms introduced by Francis [29] and later Moler and Stewart [50] are double-shift algorithms. In both cases the motivation was to create an algorithm that computes the eigenvalues of a real matrix or pencil without resorting to complex arithmetic, even though real matrices often have complex eigenvalues. The trick is to chase two complex conjugate shifts together in a single bulge.

The concept can be extended to three shifts, four shifts, or more; the size of the bulge is directly proportional to the number of shifts. Let's take a look at the double-shift case, which is the Moler–Stewart QZ algorithm. At the beginning the pencil is in Hessenberg-triangular form:

$$\begin{bmatrix} \times & \times & \times & \times & \times & \times \\ \times & \times & \times & \times & \times & \times \\ & \times & \times & \times & \times & \times \\ & & \times & \times & \times & \times \\ & & & \times & \times & \times \\ & & & & \times & \times \end{bmatrix} \begin{bmatrix} \times & \times & \times & \times & \times & \times \\ & \times & \times & \times & \times & \times \\ & & \times & \times & \times & \times \\ & & & \times & \times & \times \\ & & & & \times & \times \\ & & & & & \times \end{bmatrix}.$$

This is a Hessenberg pencil with all poles equal to ∞. The iteration begins by choosing two shifts ρ_1, ρ_2, which are either both real or a complex-conjugate pair. Then it computes the real vector

$$x = (A - \rho_1 B)B^{-1}(A - \rho_2 B)B^{-1}e_1, \qquad (7.3.1)$$

which is nonzero only in the first three entries. Then it determines a real, orthogonal Q acting on only the first three rows, such that $Qe_1 = \alpha x$ for some nonzero α. Then we left-multiply A and B by Q^*, altering rows 1, 2, and 3, to obtain

$$\begin{bmatrix} \times & \times & \times & \times & \times & \times \\ \times & \times & \times & \times & \times & \times \\ + & \times & \times & \times & \times & \times \\ & & \times & \times & \times & \times \\ & & & \times & \times & \times \\ & & & & \times & \times \end{bmatrix} \begin{bmatrix} \times & \times & \times & \times & \times & \times \\ + & \times & \times & \times & \times & \times \\ + & + & \times & \times & \times & \times \\ & & & \times & \times & \times \\ & & & & \times & \times \\ & & & & & \times \end{bmatrix}. \qquad (7.3.2)$$

Let's compare this with what happens in a pole-swapping step, which begins with a move of type I that removes the top poles σ_1 and σ_2 and replaces them by the shifts ρ_1 and ρ_2. This is done by building a unitary transformation Q using a vector x as computed by (7.2.1). In our current Hessenberg-triangular scenario, the top two poles are $\sigma_1 = \infty$ and $\sigma_2 = \infty$. If we substitute these values into (7.2.1) (after normalizing appropriately), we find that we get exactly (7.3.1). This shows that the beginning of the Moler–Stewart QZ step is exactly a pole-swapping move of type I. The result is (7.3.2), which is a block Hessenberg pencil with a 2×2 block (outlined) bearing the poles ρ_1 and ρ_2. The remaining blocks are 1×1 with poles ∞.

The remainder of the QZ step consists of returning the pair to Hessenberg-triangular form by chasing away the bulge entries shown in red in (7.3.2). Before we continue, let's enlarge the outlined region to include one of the infinite poles:

$$\begin{bmatrix} \times & \times & \times & \times & \times & \times \\ \times & \times & \times & \times & \times & \times \\ + & \times & \times & \times & \times & \times \\ & & \times & \times & \times & \times \\ & & & \times & \times & \times \\ & & & & \times & \times \end{bmatrix} \begin{bmatrix} \times & \times & \times & \times & \times & \times \\ + & \times & \times & \times & \times & \times \\ + & + & \times & \times & \times & \times \\ & & & \times & \times & \times \\ & & & & \times & \times \\ & & & & & \times \end{bmatrix}. \qquad (7.3.3)$$

7.3. Bulge chasing as pole swapping

The enclosed 3×3 pencil has eigenvalues ρ_1, ρ_2, and ∞. The first step in the bulge chase is a transformation on the right that operates on columns 1, 2, and 3 and produces zeros in B in positions $(2,1)$ and $(3,1)$.[21] It also causes some fill-in in A, as shown in (7.3.4):

$$\begin{bmatrix} \times & \times & \times & \times & \times & \times \\ \times & \times & \times & \times & \times & \times \\ + & \times & \times & \times & \times & \times \\ + & + & \times & \times & \times & \times \\ & & & \times & \times & \times \\ & & & & \times & \times \end{bmatrix} \begin{bmatrix} \times & \times & \times & \times & \times & \times \\ & \times & \times & \times & \times & \times \\ & + & \times & \times & \times & \times \\ & & & \times & \times & \times \\ & & & & \times & \times \\ & & & & & \times \end{bmatrix}. \qquad (7.3.4)$$

The eigenvalues of the outlined subpencil are still ρ_1, ρ_2, and ∞, but now they have been mixed together. This is what we called the *bulge pencil* in [66, 67] and [69, Chap. 7]. It is a subpencil of what we now call the pole pencil.

The next step is to apply a transformation on the left that operates on rows 2, 3, and 4 and creates zeros in the $(3,1)$ and $(4,1)$ positions of A and creates a bit of fill-in in the fourth row of B:

$$\begin{bmatrix} \times & \times & \times & \times & \times & \times \\ \times & \times & \times & \times & \times & \times \\ & \times & \times & \times & \times & \times \\ & + & \times & \times & \times & \times \\ & & & \times & \times & \times \\ & & & & \times & \times \end{bmatrix} \begin{bmatrix} \times & \times & \times & \times & \times & \times \\ & \times & \times & \times & \times & \times \\ & + & \times & \times & \times & \times \\ & + & + & \times & \times & \times \\ & & & & \times & \times \\ & & & & & \times \end{bmatrix}. \qquad (7.3.5)$$

The eigenvalues of the enclosed region are still ρ_1, ρ_2, and ∞, but now they have been separated into two blocks once again: a 1×1 block with eigenvalue ∞ followed by a 2×2 block, which must house the eigenvalues ρ_1 and ρ_2. This means that we have just completed a pole swap (a move of type II) in which we have swapped a 2×2 block with a 1×1 block. Now let's move our window of interest down and to the right:

$$\begin{bmatrix} \times & \times & \times & \times & \times & \times \\ \times & \times & \times & \times & \times & \times \\ & \times & \times & \times & \times & \times \\ & + & \times & \times & \times & \times \\ & & & \times & \times & \times \\ & & & & \times & \times \end{bmatrix} \begin{bmatrix} \times & \times & \times & \times & \times & \times \\ & \times & \times & \times & \times & \times \\ & + & \times & \times & \times & \times \\ & + & + & \times & \times & \times \\ & & & \times & \times & \times \\ & & & & & \times \end{bmatrix}. \qquad (7.3.6)$$

The highlighted region now contains the 2×2 block housing ρ_1 and ρ_1 at the top, followed by a 1×1 block with the next infinite pole. We are now in exactly the same situation as in (7.3.3). We continue the bulge chase just as before; the next two transformations effect a swap of the two blocks.

Eventually the 2×2 block housing ρ_1 and ρ_2 arrives at the bottom of the pencil. From here the bulge-chasing algorithm continues to eliminate bulge entries to return the pencil to Hessenberg-triangular form. This last bit amounts to a move of type I, removing the poles ρ_1 and ρ_2 and replacing them by two infinite poles.

This completes our demonstration that bulge-chasing algorithms are pole-swapping algorithms. We have used the double-shift case as an illustration, but the same picture holds for bulge chases of any degree. If we wish to do a step of degree m, we get m shifts $\rho_1, \ldots, \rho_m$ and build a bulge that carries the m shifts. This is just a block containing the m shifts as poles in a block Hessenberg pencil. Each step of the bulge chase is just a move of type II in which the big block is swapped with a 1×1 block containing an infinite pole. After many such swaps, the block reaches the bottom of the pencil. Then further annihilations dissolve the block, returning

---

[21] We leave it as an exercise for the reader to fill in the details of the construction of this transformation. The obvious method also creates a zero in the $(3,2)$ position of B. This is fine, but there is a less work-intensive method that does not create that zero. This is explained in [70, pp. 539–540], for example.

the pencil to Hessenberg-triangular form. This last bit is equivalent to a move of type I, in which the poles $\rho_1, \ldots, \rho_m$ are removed and replaced by m infinite poles.

In principle one can chase arbitrarily large bulges. Chasing, say, 30 shifts in a big bulge is equivalent to doing 30 QZ steps at once. Years ago it was thought that this would be a way to improve the performance of the QZ and QR algorithms by making more efficient use of memory [8], but it was found subsequently that this idea did not work very well in practice. Numerical investigations [66, 67] showed that the problem was *shift blurring*: in a large bulge, the shifts are not represented accurately. In other words, the eigenvalues of the bulge pencil are ill conditioned, and they will not be transmitted effectively due to roundoff errors. Therefore the expected rapid convergence does not take place. As a consequence, the idea of chasing large bulges was abandoned in favor of chasing long chains of small bulges. We can carry this lesson over to pole swapping. If we try to work with large blocks, we will likely suffer from *pole blurring*; the poles will be poorly represented due to ill conditioning. We therefore recommend working with small blocks in practice, sending information in chains of small blocks rather than one large block.

Finally we mention that there are some algorithms that need to introduce two bulges at opposite ends of the pencil and chase them in opposite directions in order to preserve some particular structure [18, 19, 42, 48]. In order to do this, we need to know how to pass two bulges through each other without mixing up the information that they contain. This question was settled in [68, 69]. In light of what we now know about pole swapping, we realize that this is simply a question of swapping two blocks.

Bibliography

[1] E. Anderson et al. *LAPACK Users' Guide*. SIAM, Philadelphia, Third edition, 1999. (Cited on p. vii)

[2] W. E. Arnoldi. The principle of minimized iterations in the solution of the matrix eigenvalue problem. *Quart. Appl. Math.*, 9:17–29, 1951. (Cited on p. 35)

[3] J. L. Aurentz, T. Mach, L. Robol, R. Vandebril, and D. S. Watkins. *Core-Chasing Algorithms for the Eigenvalue Problem*. SIAM, Philadelphia, 2018. (Cited on pp. vii, 3, 5, 7, 10, 46, 47, 57)

[4] J. L. Aurentz, T. Mach, L. Robol, R. Vandebril, and D. S. Watkins. Fast and backward stable computation of roots of polynomials, part II: Backward error analysis; companion matrix and companion pencil. *SIAM J. Matrix Anal. Appl.*, 39:1245–1269, 2018. (Cited on pp. 5, 7)

[5] J. L. Aurentz, T. Mach, L. Robol, R. Vandebril, and D. S. Watkins. Fast and backward stable computation of the eigenvalues and eigenvectors of matrix polynomials. *Math. Comp.*, 88:313–347, 2019. (Cited on pp. 7, 47)

[6] J. L. Aurentz, T. Mach, R. Vandebril, and D. S. Watkins. Fast and backward stable computation of roots of polynomials. *SIAM J. Matrix Anal. Appl.*, 36:942–973, 2015. (Cited on pp. 3, 7)

[7] J. L. Aurentz, T. Mach, R. Vandebril, and D. S. Watkins. Fast and stable unitary QR algorithm. *Electron. Trans. Numer. Anal.*, 44:327–341, 2015. (Cited on p. 69)

[8] Z. Bai and J. Demmel. On a block implementation of the Hessenberg multishift QR iteration. *Internat. J. High Speed Comput.*, 1:97–112, 1989. (Cited on p. 90)

[9] Z. Bai and J. Demmel. On swapping diagonal blocks in real Schur form. *Linear Algebra Appl.*, 186:73–95, 1993. (Cited on pp. 19, 20)

[10] Z. Bai, J. Demmel, and A. McKenney. On computing condition numbers for the nonsymmetric eigenproblem. *ACM Trans. Math. Software*, 19:202–223, 1993. (Cited on pp. 19, 20)

[11] R. H. Bartels and G. W. Stewart. Solution of the equation $AX + XB = C$. *Comm. ACM*, 15:820–826, 1972. (Cited on p. 15)

[12] B. Beckermann, S. Güttel, and R. Vandebril. On the convergence of rational Ritz values. *SIAM J. Matrix Anal. Appl.*, 31(4):1740–1774, 2010. (Cited on p. 36)

[13] M. Berljafa and S. Güttel. Generalized rational Krylov decompositions with an application to rational approximation. *SIAM J. Matrix Anal. Appl.*, 36:894–916, 2015. (Cited on pp. 39, 41, 76)

[14] D. A. Bini, B. Iannazzo, and B. Meini. *Numerical Solution of Algebraic Riccati Equations*. SIAM, 2011. (Cited on p. 16)

[15] A. Björck. *Numerical Methods in Matrix Computations*. Springer, 2015. (Cited on p. vii)

[16] K. Braman, R. Byers, and R. Mathias. The multishift QR algorithm. Part I: Maintaining well-focused shifts and level 3 performance. *SIAM J. Matrix Anal. Appl.*, 23:929–947, 2002. (Cited on pp. 51, 58)

[17] K. Braman, R. Byers, and R. Mathias. The multishift QR algorithm. Part II: Aggressive early deflation. *SIAM J. Matrix Anal. Appl.*, 23:948–973, 2002. (Cited on p. 53)

[18] R. Byers. *Hamiltonian and Symplectic Algorithms for the Algebraic Riccati Equation*. PhD thesis, Cornell University, 1983. (Cited on pp. 56, 90)

[19] R. Byers. A Hamiltonian QR algorithm. *SIAM J. Sci. Stat. Comput.*, 7:212–229, 1986. (Cited on pp. 56, 90)

[20] D. Camps. *Pole swapping methods for the eigenvalue problem: Rational QR algorithms*. PhD thesis, KU Leuven, 2019. (Cited on pp. 51, 76)

[21] D. Camps, T. Mach, R. Vandebril, and D. S. Watkins. On pole-swapping algorithms for the eigenvalue problem. *Electron. Trans. Numer. Anal.*, 52:480–508, 2020. (Cited on pp. 7, 25, 26, 41, 60)

[22] D. Camps, T. Mach, R. Vandebril, and D. S. Watkins. The RQR algorithm. 2024. arXiv:2411.17671, submitted for publication. (Cited on p. 72)

[23] D. Camps, N. Mastronardi, R. Vandebril, and P. Van Dooren. Swapping 2×2 blocks in the Schur and generalized Schur form. *J. Comput. Appl. Math.*, 373:112274, 2020. (Cited on pp. 19, 20, 29)

[24] D. Camps, K. Meerbergen, and R. Vandebril. An implicit filter for rational Krylov using core transformations. *Linear Algebra Appl.*, 561:113–140, 2019. (Cited on p. 76)

[25] D. Camps, K. Meerbergen, and R. Vandebril. A rational QZ method. *SIAM J. Matrix Anal. Appl.*, 40:943–972, 2019. (Cited on pp. 41, 42, 43, 44, 47, 51, 59)

[26] K. Dackland and B. Kågström. Blocked algorithms and software for reduction of a regular matrix pair to generalized Schur form. *ACM Trans. Math. Software*, 25(4):425–454, December 1999. (Cited on p. 44)

[27] G. De Samblanx, K. Meerbergen, and A. Bultheel. The implicit application of a rational filter in the RKS method. *BIT Numer. Math.*, 37(4):925–947, 1997. (Cited on p. 76)

[28] J. W. Demmel. *Applied Numerical Linear Algebra*. SIAM, Philadelphia, 1997. (Cited on p. vii)

[29] J. G. F. Francis. The QR transformation, part II. *Computer J.*, 4:332–345, 1961. (Cited on pp. vii, 3, 88)

[30] G. H. Golub and C. F. Van Loan. *Matrix Computations*. Johns Hopkins University Press, Baltimore, Fourth edition, 2013. (Cited on pp. vii, 44)

[31] G. H. Golub and J. H. Wilkinson. Ill-conditioned eigensystems and the computation of the Jordan canonical form. *SIAM Rev.*, 18:578–619, 1976. (Cited on p. 19)

[32] W. B. Gragg. The QR algorithm for unitary Hessenberg matrices. *J. Comput. Appl. Math.*, 16:1–8, 1986. (Cited on p. 69)

[33] R. Granat, B. Kågström, and D. Kressner. A novel parallel QR algorithm for hybrid distributed memory HPC systems. *SIAM J. Sci. Comput.*, 32:2345–2378, 2010. (Cited on p. 51)

[34] S. Güttel. Rational Krylov approximation of matrix functions: Numerical methods and optimal pole selection. *GAMM-Mitteilungen*, 36:8–31, 2013. (Cited on pp. 38, 40)

[35] N. J. Higham. *Accuracy and Stability of Numerical Algorithms*. SIAM, Philadelphia, Second edition, 2002. (Cited on pp. 18, 30)

[36] B. Kågström, D. Kressner, E.S. Quintana-Ortí, and G. Quintana-Ortí. Blocked algorithms for the reduction to Hessenberg-triangular form revisited. *BIT Numer. Math.*, 48(3):563–584, July 2008. (Cited on p. 44)

[37] B. Kågström and P. Poromaa. Computing eigenspaces with specified eigenvalues of a regular matrix pair (A, B) and condition estimation: Theory, algorithms and software. *Numer. Algorithms*, 12:369–407, 1996. (Cited on p. 29)

[38] B. Kågström and P. Poromaa. LAPACK-style algorithms and software for solving the generalized Sylvester equation and estimating the separation between regular matrix pairs. *ACM Trans. Math. Softw.*, 22:78–103, 1996. (Cited on p. 29)

[39] L. Karlsson, D. Kressner, and B. Lang. Optimally packed chains of bulges in multishift QR algorithms. *ACM Trans. Math. Software*, 40:1–15, 2014. (Cited on pp. 57, 58)

[40] D. Kleinman. On an iterative technique for Riccati equation computations. *IEEE Trans. Automat. Control*, 13(1):114–115, 1968. (Cited on p. 16)

[41] D. Kressner. On the use of larger bulges in the QR algorithm. *Electron. Trans. Numer. Anal.*, 20:50–63, 2005. (Cited on p. 58)

[42] D. Kressner, C. Schröder, and D. S. Watkins. Implicit QR algorithms for palindromic and even eigenvalue problems. *Numer. Algorithms*, 51:209–238, 2009. (Cited on pp. 56, 90)

[43] A. B. J. Kuijlaars. Which eigenvalues are found by the Lanczos method? *SIAM J. Matrix Anal. Appl.*, 22:306–321, 2000. (Cited on pp. 36, 76)

[44] A. B. J. Kuijlaars. Convergence analysis of Krylov subspace iterations with methods from potential theory. *SIAM Rev.*, 48:3–40, 2006. (Cited on pp. 36, 76)

[45] B. Lang. *Effiziente Orthogonaltransformationen bei der Eigen- und Singulärwertzerlegung*. Habilitationsschrift, Universität Wuppertal, Wuppertal, Germany, 1997. (Cited on p. 51)

[46] B. Lang. Using level 3 BLAS in rotation-based algorithms. *SIAM J. Sci. Comput.*, 19:626–634, 1998. (Cited on p. 51)

[47] R. B. Lehoucq, D. C. Sorensen, and C. Yang. *ARPACK Users' Guide: Solution of Large-Scale Eigenvalue Problems with Implicitly Restarted Arnoldi Methods*. SIAM, Philadelphia, 1998. (Cited on p. 73)

[48] T. Mach, T. Steel, R. Vandebril, and D. S. Watkins. Pole-swapping algorithms for alternating and palindromic eigenvalue problems. *Vietnam J. Math.*, 48:679–701, 2020. (Cited on pp. 56, 90)

[49] T. Mach and R. Vandebril. On deflations in extended QR algorithms. *SIAM J. Matrix Anal. Appl.*, 35(2):559–579, January 2014. (Cited on p. 7)

[50] C. B. Moler and G. W. Stewart. An algorithm for generalized matrix eigenvalue problems. *SIAM J. Numer. Anal.*, 10:241–256, 1973. (Cited on pp. 5, 44, 88)

[51] A. Ruhe. An algorithm for numerical determination of the structure of a general matrix. *BIT Numer. Math.*, 10:196–216, 1970. (Cited on p. 19)

[52] A. Ruhe. Rational Krylov sequence methods for eigenvalue computation. *Linear Algebra Appl.*, 58:391–405, 1984. (Cited on p. 38)

[53] A. Ruhe. Rational Krylov algorithms for nonsymmetric eigenvalue problems. II. Matrix pairs. *Linear Algebra Appl.*, 197–198:283–295, 1994. (Cited on p. 38)

[54] A. Ruhe. The rational Krylov algorithm for nonsymmetric eigenvalue problems. III: Complex shifts for real matrices. *BIT Numer. Math.*, 34(1):165–176, 1994. (Cited on p. 38)

[55] A. Ruhe. Rational Krylov: A practical algorithm for large sparse nonsymmetric matrix pencils. *SIAM J. Sci. Comput.*, 19:1535–1551, 1998. (Cited on p. 38)

[56] I. Schur. Über die charakteristischen Wurzeln einer linearen Substitution mit einer Anwendung auf die Theorie der Integralgleichungen. *Math. Ann.*, 66:488–510, 1909. In German. (Cited on p. 1)

[57] D. C. Sorensen. Implicit application of polynomial filters in a k-step Arnoldi method. *SIAM J. Matrix Anal. Appl.*, 13:357–385, 1992. (Cited on p. 73)

[58] T. Steel, D. Camps, K. Meerbergen, and R. Vandebril. A multishift, multipole rational QZ method with aggressive early deflation. *SIAM J. Matrix Anal. Appl.*, 42:753–774, 2021. (Cited on pp. vii, 51, 55, 76, 79, 84)

[59] G. W. Stewart. On the sensitivity of the eigenvalue problem $Ax = \lambda Bx$. *SIAM J. Numer. Anal.*, 9:669–686, 1972. (Cited on pp. 5, 16)

[60] G. W. Stewart. Error and perturbation bounds for subspaces associated with certain eigenvalue problems. *SIAM Rev.*, 15:727–764, 1973. (Cited on p. 16)

[61] G. W. Stewart. Algorithm 506: HQR3 and EXCHNG. Fortran subroutines for calculating and ordering the eigenvalues of a real upper Hessenberg matrix. *ACM Trans. Math. Software*, 2:275–280, 1976. (Cited on pp. 19, 20)

[62] G. W. Stewart. A Krylov–Schur algorithm for large eigenproblems. *SIAM J. Matrix Anal. Appl.*, 23:601–614, 2001. (Cited on p. 76)

[63] L. N. Trefethen and D. Bau, III. *Numerical Linear Algebra*. SIAM, Philadelphia, 1997. (Cited on p. vii)

[64] P. Van Dooren. A generalized eigenvalue approach for solving Riccati equations. *SIAM J. Sci. Stat. Comput.*, 2:121–135, 1981. (Cited on pp. 19, 25, 26, 29)

[65] D. S. Watkins. Bidirectional chasing algorithms for the eigenvalue problem. *SIAM J. Matrix Anal. Appl.*, 14(1):166–179, 1993. (Cited on p. 52)

[66] D. S. Watkins. Forward stability and transmission of shifts in the QR algorithm. *SIAM J. Matrix Anal. Appl.*, 16:469–487, 1995. (Cited on pp. viii, 56, 89, 90)

[67] D. S. Watkins. The transmission of shifts and shift blurring in the QR algorithm. *Linear Algebra Appl.*, 241–243:877–896, 1996. (Cited on pp. viii, 56, 81, 89, 90)

[68] D. S. Watkins. Bulge exchanges in algorithms of QR type. *SIAM J. Matrix Anal. Appl.*, 19:1074–1096, 1998. (Cited on pp. viii, 56, 90)

[69] D. S. Watkins. *The Matrix Eigenvalue Problem: GR and Krylov Subspace Methods*. SIAM, Philadelphia, 2007. (Cited on pp. vii, 15, 16, 33, 44, 47, 56, 89, 90)

[70] D. S. Watkins. *Fundamentals of Matrix Computations*. Wiley, New York, Third edition, 2010. (Cited on pp. vii, 2, 3, 5, 7, 33, 36, 37, 44, 45, 46, 47, 73, 74, 89)

[71] D. S. Watkins. Francis's algorithm. *Amer. Math. Monthly*, 118:387–403, 2011. (Cited on pp. vii, 3, 33, 46, 47)

[72] J. H. Wilkinson. *The Algebraic Eigenvalue Problem*. Clarendon Press, Oxford University, 1965. (Cited on p. 43)

Index

aggressive early deflation, 53
   pencil case, 54
anti-identity matrix, 87
Arnoldi configuration, 75
Arnoldi process, 35–36
   generalized, 37, 76
   implicitly restarted, 73, 75
   rational, 38, 76

basic algorithm, 44
   convergence, 47
   variations, 51
bulge pencil, 56, 81
   initial, 81
   pole, *see* pole pencil

column space, 3
continuation vector, 37, 38, 76
convergence, 46, 84
   of basic algorithm, 47, 59
   of QZ algorithm, 47
core transformation, 7
   double-arrow notation, 8

deflating pair, 5

eigenvalue, 1, 4
   generalized, 4
   infinite, 4
eigenvector, 1, 4
   generalized, 4
equivalence, 5
Euclidean norm, 7

filtering, 73, 74, 76
flip matrix, 87
floating-point arithmetic, 2
flop, 2
Francis's algorithm, 3, 36, 74
fusion, 9

Givens rotation, *see* rotation

Gram–Schmidt process, 35

Hessenberg matrix, 2
   unitary, 11, 67
Hessenberg pair, 39, 41
   proper, 42
Hessenberg pencil, 41
   reduction to, 44

implicit restarts, 73
inner product, 6
invariant subspace, 3, 35

Kronecker product, 20
Krylov process, 73
   Arnoldi, *see* Arnoldi process
   rational, 38, 60, 76
Krylov subspace, 33
   rational, 60
Krylov–Schur algorithm, 76

matrix
   anti-identity, 87
   flip, 87
   Hessenberg, 2
   sparse, 33
   triangular, 1
   unitary, 1, 67
matrix pencil, 5
Moler–Stewart algorithm, *see* QZ algorithm
move
   type I, 41, 69, 80
   type II, 43, 69, 79

Newton's method, 16, 24
Newton–Kleinman iteration, 16
norm, 7

optimally packed shifts, 57
orthogonal vectors, 6
orthonormal vectors, 7

pair
   Hessenberg, 5, 39, 41
   Hessenberg-triangular, 5
   proper Hessenberg, 42
pencil, 5
   block-triangular, 21
   Hessenberg, 5, 41
   Hessenberg-triangular, 5
   pole, 41
   regular, 5
   singular, 5
pertranspose, 27, 30, 87
pole, 38, 41
   Rayleigh quotient, 45
pole pencil, 41
poles, 39
proper Hessenberg pair, 42

QR algorithm, *see* Francis's algorithm
QZ algorithm, 5, 45
   convergence, 47

range, 3
rational Krylov process, 38
rational Krylov subspace, 60
Rayleigh quotient pole, 45
Riccati equation, 16
   generalized, 24
rotation, 7, 11, 17
RQR algorithm, 69
   performance, 72
RQZ algorithm, 44

Schur's theorem, 2
   generalized, 5, 29
   Schur decomposition, 2
shift, 44, 45, 47, 74, 83
   multiple shifts, 51, 53
   optimally packed, 57
   Rayleigh-quotient, 44
shift-and-invert strategy, 37, 76

shift-through operation, 10
similarity, 1
similarity transformation, 1
    partial, 36
sparse matrix, 33
spectral norm, 7
spectrum, 1, 4
subspace

deflating pair, 5
    invariant, 3
Krylov, 33
    rational Krylov, 60
Sylvester equation, 15
    generalized, 22
    unique solution, 20

tensor product, 20

triangular matrix, 1
    swapping eigenvalues, 17
turnover operation, 10

unitary Hessenberg matrix, 67
unitary matrix, 1, 67

vec(X), 20